Benjamin Samuel Williams

The Orchid Grower's Manual

Containing brief Descriptions of upwards of four hundred and forty of orchidaceous

Plants

Benjamin Samuel Williams

The Orchid Grower's Manual
Containing brief Descriptions of upwards of four hundred and forty of orchidaceous Plants

ISBN/EAN: 9783337106713

Printed in Europe, USA, Canada, Australia, Japan

Cover: Foto ©ninafisch / pixelio.de

More available books at **www.hansebooks.com**

THE

ORCHID-GROWER'S MANUAL

CONTAINING BRIEF DESCRIPTIONS

OF

UPWARDS OF FOUR HUNDRED AND FORTY ORCHIDACEOUS PLANTS,

TOGETHER WITH

NOTICES OF THEIR TIMES OF FLOWERING AND MOST APPROVED MODES OF TREATMENT;
ALSO, PLAIN AND PRACTICAL INSTRUCTIONS RELATING TO THE GENERAL
CULTURE OF ORCHIDS; AND REMARKS ON THE HEAT,
MOISTURE, SOIL, AND SEASONS OF GROWTH
AND REST BEST SUITED TO THE
SEVERAL SPECIES.

BY

BENJAMIN SAMUEL WILLIAMS, F.R.H.S.,
PARADISE NURSERY, HOLLOWAY, LONDON, N.;
AUTHOR OF "HINTS ON THE CULTIVATION OF FERNS," &c., &c.

SECOND EDITION.

LONDON:
CHAPMAN AND HALL, 193, PICCADILLY.
1862.

TO

CHARLES BORHAM WARNER, Esq.,

FELLOW OF THE ROYAL HORTICULTURAL SOCIETY OF LONDON,

The following Pages

ARE RESPECTFULLY INSCRIBED

BY

HIS OBEDIENT SERVANT,

B. S. WILLIAMS.

PREFACE TO THE SECOND EDITION.

It is just ten years since I first published the "Orchid Grower's Manual," to which the remarks in the preface to the first edition relate. At that time it contained an enumeration of all the species known to me to be good and showy. In the following pages I have added many new and beautiful introductions that have been made to Orchids generally within the last few years. I may take this opportunity also of saying, that in bringing a Second Edition of this work before the public, it gives me great pleasure to know that the first issue has been the means of promoting and increasing the cultivation of this glorious class of plants, and that I have had ample assurances that some hundreds of cultivators have found my efforts useful. Indeed, repeated solicitations from all parts of the country have chiefly induced me to bring out a fresh edition, which I hope will be found even more valuable than the first. I have endeavoured to make it as plain as possible, in order that persons even with limited acquaintance of the management of Orchids might understand it. More than fifty pages of new matter have been added, containing minute details as to all the more important wants of cultivators. I have also carefully described all new varieties proved by me to be worth growing—whether for purposes of exhibition or for those of private cultivation. Curiosities of mere botanical interest have been omitted, my study having been to give a good list of such kinds as have handsome flowers, or are otherwise ornamental in the way of fine foliage; let me hope, therefore, that the present edition may be received with as much approbation and prove even more satisfactory than the first.

B. S. WILLIAMS.

Paradise Nursery, Holloway, N.

PREFACE TO THE FIRST EDITION.

The following papers were written nearly four years ago, at the suggestion of Mr. Bellenden Ker, who had commenced the cultivation of Orchids, and whose gardener then employed was not conversant with their treatment. At first it was intended that the observations should have been printed in a small volume; but the papers were afterwards offered to the Editor of the "Gardeners' Chronicle," in which publication they appeared under the title of "Orchids for the Million." They were preceded by the following remarks, made by the gentleman at whose suggestion they were written.

"Being desirous of growing a few of the more showy Orchidaceous plants, but as neither myself nor my then gardener was acquainted with the mode of their cultivation, I applied for some information on this head to Mr. Williams, the gardener of my neighbour, Mr. Warner, and who kindly, from time to time, gave me such instruction as was necessary for a beginner. There is a notion amongst gardeners that the cultivation of these plants is attended with great difficulty; that different houses for different climates are necessary; and that the secret of good culture is only known to few; that, in fact, no one but those having a large establishment, and a gardener expressly skilled in Orchidaceous culture, should attempt to grow them. Mr. Bateman, in the preface to his great work, intimates that their cultivation is to be left to the aristocratic, whilst the more humble florist is to be confined to his Carnations, Auriculas, Dahlias, and such-like flowers. Mr. Williams' plan seemed to me very simple, and also that one house might easily be made to answer the purpose, if properly managed; at least for those who are not ambitious of possessing a very large collection, or of growing such as are most impatient of culture. Having derived much advantage from the instruction afforded to me, I recommended Mr. Williams to make notes of his mode of culture, and since these were written I have advised him to publish them. I trust that, to humble beginners like myself, I have done good service by this advice; and I cannot but think that ere long many will be induced to attempt the cultivation of this beautiful tribe of plants, who, for want of plain instructions, have hitherto been deterred from making a beginning. A very small house is sufficient, hot-water pipes or tanks are now cheap, and a small boiler may be had for 2*l.* 10*s.*, or less; and glass (thanks to Sir Robert Peel) is also within the means of the humblest gardener; and those who refer to Mr. Rivers' account in the *Chronicle* of how to make cheap lights, and build cheap houses, will find that for 40*l.*, or less, a very sufficient Orchid-house, with hot-water apparatus, may be put up.

"Knowing the pleasure derived by many in the humblest classes from the cultivation of flowers, and how much talent, patience, and ingenuity are often displayed by such amateurs, I feel gratified by the hope that I may possibly be the means of increasing the harmless amusement of many. It is not likely that Mr. Williams' instructions will interest or inform those cultivators — Mr. Mylam, Mr. Blake, Mr. Bassett, Messrs. Veitch, Rollisson, &c., or others at the head of the great collections of this country; it is merely for the benefit of the beginner that they are intended. As regards the fitness of Mr. Williams to instruct, the best proof I can offer of this is, that for some years he has exhibited, both at Chiswick and in the Regent's Park, and a reference to the prize lists will show that he was always successful, and, during the past years (*i. e.*, previous to 1848), obtained 23 gold and 12 silver medals for Orchids, and 14 silver ones for Ferns. Mr. Warner's collection not being so large as those of many others, it was only the last two years that Mr. Williams competed in the highest class of twenty plants."

Since the first appearance of these observations, I have revised them, and have added notices of several desirable, showy, and new species; and the following pages will contain short and plain, but, it is believed, accurate descriptions of more than 260 of the best Orchids now in cultivation. I cannot resist this opportunity of expressing my gratification at finding that it has come to my knowledge that the publication of these papers has already induced several persons to commence the cultivation of this interesting class of plants.

B. S. W.

THE ORCHID-GROWER'S MANUAL.

INTRODUCTION.

The nature and habits of Orchidaceous plants are now better known than they were some few years ago, and we have become better acquainted with the conditions under which they grow in the countries where they are indigenous. There they are exposed to a dry season, during which they are at rest; and to a rainy season, when the heat is high, and the air moist nearly to saturation. To grow Orchids in perfection, their native climate must, to a certain extent, be imitated; viz. they must have a period of rest in a dry and comparatively cool atmosphere, and during their growth and flowering they should be exposed to a high moist atmosphere; but as they principally grow on the trunks and branches of trees, it is important that they should be exposed to a free current of air, and also to light: this is essential, except in the case of some few species which require shading, to prevent the plants being exposed to the direct action of the sun's rays, which are apt to scorch the leaves. Great heat and moisture are only necessary while the plants are in vigorous growth, and this period should be during spring and summer, the best season of rest being from about November till February: and it is the long period of rest which predisposes the plants to blossom. Of course, rules as regards growth and rest can only be stated in general terms. There are certain kinds which are never entirely inactive throughout the whole year. And again, even of those which do go to rest periodically, on the completion of their growth, it does not always happen that their dormant season corresponds with that at which the largest numbers go to rest.

Directions for the Period of Growth.

When the growing season commences, raise the temperature of the East India house, or house in which the greatest heat is maintained, to 65° by night and 70° by day ; by sun heat it may be allowed to range to 75°, and as the days lengthen, so the temperature may be permitted to increase ; and during the months of May, June, July, and August it should range from 70° to 75° by night, and from 75° to 80° by day, and by sun heat to 85°, or even 90° : this will not do any harm, provided the plants are shaded from the direct rays of the sun. The Mexican or cooler house should be 60° by night, and from 65° to 70° by day ; and as the days lengthen, so the temperature may be allowed to increase ; and during the months of May, June, July, and August the night heat may range from 65° to 70°, and by day from 70° to 85°. Great attention should, at the same time, be paid to the state of the atmosphere, as regards moisture ; at all times of the year this is of much importance to the successful growth of the plants, for they derive the greater part of their subsistence from moisture in the air, so that wherever any plants are growing, the atmosphere should be well supplied with moisture : this is obtained by pouring water over the tables, walls, and paths of the house every morning and afternoon, and by keeping the hot-water tanks full ; this will cause a nice gentle steam to rise, which is of great value whilst the plants are in a vigorous state of growth, especially as regards the East Indian Orchids, such as *Aerides*,

Saccolabiums, Vandas, Phalænopsis, Dendrobiums, and many others requiring a high temperature, with a considerable degree of moisture. The Mexican Orchids, most of which come from a cooler climate, not so saturated with water, of course require less heat and moisture, but these should also have a considerable degree of warmth during their growing season.

Water.

This should be administered with great care, especially in the case of plants just starting into growth, as, if watered too profusely, the young shoots are apt to be affected by the moisture of the house, and liable to what is termed damping off; whilst, therefore, the shoots are young, only enough of water should be given to keep the peat moist in which the plants are grown. As they advance in growth, more may be given; and when the pseudo-bulbs are about half grown, they may have a good supply at the roots. My practice is to shut up the house in the spring of the year about three o'clock; and in May, June, July, August, and September I shut it up about an hour later, when the heat of the sun is on the decline. I then usually give a gentle syringing with water as nearly as may be of the same temperature as that of the house. In fine weather, the temperature from sun heat will rise frequently as high as 95°, or even more; but I have never experienced any injury from this, so long as the house was saturated with moisture, in which case there is no fear of injury to the plants. The house should be dried up once a day, if possible, by means of ventilation. In syringing, be careful not to wet the young shoots too much. The syringe should be furnished with a fine rose, so as to cause the water to fall on the plants in imitation of a gentle shower of small rain; but this syringing should only be done after a hot summer's day. Those plants which are growing on blocks of woods should be syringed twice a-day in the summer time; and I also find it a good method, during the growing season, to take the blocks down, and dip them in water till the wood and moss are thoroughly soaked. Plants in baskets should likewise be taken down and examined, and if they are dry, they should be soaked in the water. This is also a good mode of getting rid of many insects that harbour in the moss, such as the woodlouse and cockroach; when the moss is soaked, they will come to the top, and then they may be easily killed. Rain or pond water is the best.

On the Cultivation of Tropical Orchids.

Among Orchids some are termed terrestrial, that is, they grow in earth; the genera *Phajus, Calanthe, Bletia, Cyrtopodium, Cypripedium,* &c., all draw support from the ground. *Epiphytes,* the other class, inhabit trees and rocks, from which, however, they derive little or no nourishment. These are by far the most numerous and interesting. They are found adhering to the arms of living trees, and some of them delight in elevated situations in lofty forests. Others, again, grow upon low trees, some on rocks and mountains, some on trees overhanging rivers, and some near dripping rocks. The latter, of course, require a particularly damp atmosphere to grow in; others are found in dense woods, where scarcely any sun can penetrate; these like a shady moist atmosphere, whilst those in more elevated situations do not need so much shade as the last. A knowledge of the different habitats of the various species is essential to the careful grower, in order that he may, as far as his means permit, place them in circumstances similar to those in which they make their natural growth; and it is, perhaps, to inattention on this point that the want of success in the culture of some Orchidaceous plants, by even the most successful of our cultivators, is to be attributed.

Mode of Potting and the Materials to be used.

When the season of rest is over, many kinds will require re-potting, but I have not confined my practice to that time only; no season can be determined on absolutely as the proper one for this operation. The months of February and March are the best times to pot some of them, that is, after the resting season. Those that do not require potting should be top-dressed with good fibrous peat, removing the old soil from the top without breaking the roots of the plants. This also affords the means of getting rid of many insects which harbour in the old soil. The pots should be thoroughly cleansed from the mould, moss, and dirt, too often seen covering those in which Orchids are growing. Previously to potting the plants, they should not receive any water for four or five days. Some, however, should be potted at a period somewhat later, viz., just as they begin to grow. All the species of *Phajus, Calanthe, Dendrobium, Stanhopea, Cyrtopodium, Brassia, Miltonia, Sobralia, Bletia, Oncidium*, and many others, require this treatment. *Lælias, Cattleyas, Saccolabiums, Aerides, Vandas*, and similar plants, should be potted just before the commencement of their growing season. The chief point to be attended to in all potting is that the pots be well drained; the best material for drainage is potsherds or charcoal. Before potting, be particular to have the pots perfectly clean inside and out, and the broken potsherds should be washed: after this is done, select a pot according to the size of the plant; do not give too much pot room. Some plants require shifting once a year, while it may not be necessary to shift others oftener than once in two or three years; but if a plant becomes sickly or soddened with wet, the best way to bring it into a healthy state is to turn it out of the pot or basket, and wash the roots carefully with some clean water, cutting off such of the fibres as are dead, then to re-pot it, not giving it much water till it begins to make fresh roots. The best pots are those in ordinary use. Some employ slate pots, but they are not, in my opinion, so good for Orchids as those made of clay.

In potting large plants there should be a small pot turned upside down in the bottom of the large one, then fill in with potsherds or charcoal broken up into pieces, about two inches square for large plants; smaller plants should not have pieces so large; then introduce potsherds till within three or four inches of the rim, and afterwards put on a layer of moss to prevent the peat from impeding the drainage, and to let the water pass off quickly. This is of great importance, for if not attended to the water will become stagnant, and the soil sodden, which is fatal to the growth of the plant. The grand point to be observed in the successful culture of Orchids, as well as of other plants, is good drainage; without that it is hopeless to try to keep the plants long in a healthy condition. The best material for potting the different kinds of epiphytes in is good rough fibrous peat and sphagnum moss; after the layer of moss is applied, then fill up with peat. This should be broken into lumps about the size of a hen's egg: I always use broken potsherds or charcoal mixed with the peat. The plant should be elevated above the rim of the pot two or three inches, taking care to have all the pseudo-bulbs above the soil; then put some peat on the top of the roots so as to cover them, employing a few small pegs to keep the soil firmly in the pot. After the plants are potted I fix a stick in the centre of the plant to keep it firm. In shifting, carefully shake off all the old soil you can without injuring the roots, and be careful not to give too much water at first; but after the plants begin to make more root, they may have a good supply. The best material for those in baskets is sphagnum and broken potsherds. The basket should suit the size of the plant; but do not have it too large, for it will not last more than two or three years, at which time, probably, the plant will require shifting into a larger one. There should be placed a layer of moss at the bottom of the basket, then a few potsherds, then fill up with moss and potsherds mixed. Take the plants carefully out of the old basket without breaking the roots, shake off all the old moss, place the plant on the new material, about level with the top of the basket; put a stick in the centre, to keep it firm, and finish by giving a gentle watering.

Those plants that require wood to grow upon should have live moss attached to the blocks, if by experience they are found to require it; some, however, do better on bare blocks, but they need more moisture, as they are then entirely dependent on what is obtained from the atmosphere. In fastening them firmly on the blocks, have some copper nails and drive them into the block: then, with copper wire, secure the plants firmly to the wood. As soon as they make fresh roots they will cling to the block, and the wire may be taken away.

Material for Terrestrial Orchids.

These require a stronger compost than epiphytal kinds. They should be potted just when they begin to grow, after the resting season; they do not need so much drainage as epiphytes. The compost I use for them is turfy loam chopped into pieces about the size of a walnut, leaf-mould or peat, and a little rotten cow or horse-dung; these should be all mixed well together. The plants require a good-sized pot; put about two inches of drainage at the bottom, on that a layer of moss, then some rough peat, and finish with the compost just mentioned; place the plant one inch below the rim of the pot; water sparingly at first, but when the young growths are about six inches high, they may have a good supply.

Treatment of Fresh-imported Plants.

When unpacked these should be sponged over—every leaf and bulb—and all the old decayed parts removed. There are many insects that harbour about them, such as the cockroach, and different kinds of scale, which are great pests. When clean they should be laid on dry moss and placed in some shady part of the house where it is rather cool and dry. Too much light, heat, and moisture at first is injurious to them. The moss should be gradually moistened, and when they begin to grow and make roots they should be potted or put on blocks or in baskets, but care should be taken not to have the pots too large, over-potting being dangerous.

As soon as they begin to grow, those which come from the hotter parts of India should be put at the warmest end of the house, but they should not have too much moisture at first: those which come from the more temperate regions should be kept in the coolest part of the house, and they should not be allowed to stand under drip, as this frequently rots the young shoots as soon as they appear. Such plants as *Vandas, Saccolabiums, Aerides, Angræcums, Phalænopsis*, are fastened on blocks as soon as they are received, and I place them so that the plants hang downwards, in order that no water may lodge about them, till they begin to grow and form new roots: this is much the safest mode of treating these valuable Orchids.

ADVICE TO COLLECTORS.

THERE are many different ways of importing Orchids to this country. I have seen some that have arrived in good condition while others have been completely destroyed by not being properly prepared before starting. The first and most important thing is to send the plants away at the proper time; the next thing is to prepare them for their journey. My opinion is, that the plants should be sent away from their native country during the dry season, which is when they are at rest. While inactive their leaves contain little sap; but if sent away when they are growing, the foliage is tender and in danger of being bruised, a circumstance which accelerates decay on their journey; another fact in favour of dormant importation is that, if sent in a growing state or just as they are starting into growth, the young shoots come weak and dwindling, and very often die outright as soon as exposed to light. I have seen many a fine mass of *Cattleyas* with all their leading growths completely rotten, of course lessening the plants in value compared with such as arrive perfect and are just ready to start into growth as soon as they get into a warm house. Plants with pushing pseudo-bulbs are also apt to lose their leading eyes, an accident fatal to some Orchids, for many do not break well from old bulbs. *Aerides, Saccolabiums, Vandas, Angræcums*, and similar plants that have no fleshy bulbs to support are best imported after they have become established on flat pieces of wood, so that they can be nailed to the sides of their travelling cases. I received some from Manilla last spring, including *Phalænopsis Schilleriana, P. rosea, P. amabilis, P. Lobbi, Aerides quinquevulnerum,* and *Vanda violacea*, all established and sent off in the way just described; these had evidently been growing some time before starting for this country, for their roots firmly adhered to the wood and many of their leaves were as green as though they had been in an Orchid House instead of a glass-topped case. One point of importance is to take care to well secure the plants to the sides of the cases; because, if allowed to roll about, they get bruised and soon rot, which is very vexing after all the trouble and expense bestowed in importing them. In the case of *Phalænopsis Schilleriana*, received last spring, some of the pieces of wood had become loose, rolling about during the journey and causing injuries; if, therefore, you find a leaf bruised, the best way is to cut it off at once before decay begins; for if allowed to go on, there will be danger of the whole plant being destroyed. With the cases just alluded to I had also a close box filled with *Phalænopsis* packed in the dry bark of trees, which I consider a bad material for such tender-leaved plants; when I unpacked this box there was not a green leaf to be seen, the shaking of the long journey, combined with the rough material just named, had destroyed all the foliage. If these had been packed in very dry soft moss they would most likely have come safe. I have received plants in good condition from India in close boxes packed in dry soft shavings, while on the other hand I have also seen many spoiled in that way. The cause of failure I attribute to their not being packed in a proper state; the plants themselves, as well as the material employed, should be well dried before packing, and care should be taken to avoid bruises, which often prove fatal. *Cattleyas* and plants with similar pseudo-bulbs I have received in close boxes from the Brazils packed in dry shavings, and have found them, when opened, in good condition; but care had evidently been taken to pack them firmly in the boxes, so that they did not roll about on their journey.

The best time of year for receiving Orchids in this country is, if possible, the spring, in order that they may have the summer before them to get established.

With *Anæctochili* the best way is to tie some moss round their roots and stems to keep them firm, leaving the foliage just above the moss, and they should occupy a small case by themselves: these little things are very tender, therefore they require a great deal of care to keep them right. On arrival, pot them in some dry soil (see *Anæctochili*), and put them in a close place with little heat at first, until they begin to grow; afterwards pot them in small pots, and place them under hand-glasses or in a frame, giving them the treatment usually recommended for this class of plant.

Cases in which Orchids are sent home ought to be made strong, and roofed with good stout glass not easily broken; for I have often seen plants spoiled by the glass being fractured. Through an accident of this kind, salt water and cold air get in, both of which are very injurious. All cases should be air and water-tight; and to prevent the glass being broken, the best thing to place over it is some strong iron wire; the sash bars ought also to be made very strong: and the case must not be too near heated surfaces or fires in the vessel. I have seen many boxes of plants spoiled by positions of that kind, the leaves being completely dried up. They ought to be placed in as warm a situation as you can, but by no means near any fires.

ORCHID HOUSES.

A HOUSE most suitable for the culture of Orchids is the first thing to be considered. In my opinion the best houses are those with span roofs facing east and west. They should not be more than ten or eleven feet high in the centre, seventeen or eighteen feet wide, and about sixty feet long, with a glass division in the centre to separate them into two houses—one for plants that come from the East Indies, which ought to be next the boiler, the other for those that come from cooler regions. There ought to be upright sashes on both sides of the house with glass from twelve to fifteen inches high, but not to open. Many Orchid growers object to side sashes and generally recommend brickwork up to the spring of the roof: but that is not, in my opinion, the best plan; on the contrary, I would advise any one about to build an Orchid house to have upright sashes on both sides and at each end. From experience I have found that Orchids cannot have too much light, which is the only way to get good strong ripe pseudo-bulbs fit for flowering. Small houses are best : in different parts of the country there exist large houses, but in no instance have I seen plants growing well in them; such houses take great heat to keep them at the proper night temperature, and after all, they seem ill adapted to the wants of the plants. I would advise all large Orchid houses to be pulled down or turned to other purposes, and their places occupied by small ones ; the expense of the operation would soon be saved in the reduction that would take place in the cost of pipe and fuel.

The house at this place is of the size recommended above, and no Orchid house could answer better. It has been built about seven years, and is well worth inspection. It affords plenty of room for the plants to show themselves to advantage, and it has likewise roomy paths, which is a great recommendation ; for nothing is more unpleasant than not being able to inspect the plants with comfort. The inside dimensions of my house are as follows :—it is sixty feet long, eleven feet high in the centre, with a glass division, making as it were two houses. The width is eighteen feet ; there is a table six feet wide up the centre, and a path all round three feet wide ; there are side-tables three feet wide, covered with slate. The floor is concreted, three inches thick, and then covered with Portland cement, which forms a capital surface. The whole is heated with hot water, in three rows of four-inch pipes on each side, having valves to stop or turn-on the water as required. The boiler is an improved saddle, which answers well, and when properly set, as mine is, not easily beaten. This same boiler works two other large houses, one of which is fifty feet, and the other seventy feet long, and all are kept at stove heat. On both sides of this house are upright sashes, as recommended above. It is glazed in the same way as is recorded in my remarks on glazing (see next page), and it has four ventilators on each side close to the hot-water pipes in the brickwork, and one at the end over the doorway. The top lights are fixed.

Heating.

With respect to this, nothing is better than hot water in four-inch pipes; it is also better to have plenty of piping than to have too little; there is nothing saved by economy of that kind; better spend a little more money for material at first than have to add afterwards; by having plenty of pipe you do not require so much fire-heat, which is better for the plants, and you save the expense in fuel in a very short time. I should, therefore, advise four pipes instead of three; by having four, you do not require to drive the fire so much on a frosty night. I never use steam; I find that I can get plenty of moisture without it, by pouring water on the tables and paths, which I consider much better than so much steam poured on the plants direct from rusty pipes or tanks. The boiler should be outside of the house. I have seen them set underneath, which I think bad. I remember going to see a collection of Orchids where the boiler was so situated. The gentleman said he had put it there to economise heat, and the plants were growing very finely at the time. I remarked that I should be afraid of smoke getting into the house; he replied, there is no fear of that, for I have got the boiler well covered over. Only a few months afterwards I went to see the same collection, the smoke had got into the house, and had spoiled many of his plants. I merely mention this to show the ill effects of a boiler being set so that smoke can get into the house. When outside there is no fear of such a disaster.

Glazing.

The laps here are very close; the squares are two feet six inches long by nine inches in breadth; the glass in the upright sashes at the sides is fourteen inches long by nine inches wide; twenty-one oz. glass is best, not being easily broken. I remember seeing an Orchid house after a hail-storm much injured, a large portion of its valuable contents being nearly spoiled. The glass used in this instance was only sixteen oz., while, if it had been twenty-one oz., it would most likely have withstood the storm. I should recommend twenty-one oz. glass, or even more, as not being likely to get broken by cleaning or otherwise. Too large squares are bad, being apt to get broken by frost. The upright glass at the sides ought to be of the size stated above, to correspond with that in the roof and also in the ends and the door; the sash bars should have a small groove down them to carry drip to the bottom, in order to prevent it dropping on the plants; even small pieces of zinc nailed to the bars serve to form a sort of gutter to carry off drip.

Ventilation.

This is of great importance; for if cold air is allowed to pass among the plants they will not thrive, and all care which has previously been bestowed on them will be in vain. Means of ventilation should, therefore, be provided for, near the ground, close to the hot-water pipes, in order that the air may be warmed in entering the house. In the houses here there are four ventilators on each side, two to each house. The ventilating shutters are made of wood, about two feet long, one foot wide; sliding slate ventilators answer perhaps better: there should be one glass ventilator at the south end near the roof, and one at the north end, as the top lights of the house are best not made to slide.

Shading.

Every Orchid-house requires to be shaded. The best material is canvas, and there should be blinds on each side, with a strong lath at the top to nail the canvas to, and a roller at the bottom. The canvas must be nailed to the roller, but care should be taken in doing this that the awning rolls up regularly from bottom to top. I never, however, allow the canvas to be down, except when the sun is powerful, for I find that too much shade is injurious to most kinds of Orchids. The awning will also be useful in the winter season for covering the house during a frosty night, being a great protection to those plants that are near the glass. It is advisable to have a covering on the top of the house for the protection of the canvas when rolled up, in order to keep it from wet.

Cisterns.

Slate cisterns, for collecting water which falls on the roof, are very important in a house. Cisterns on each side over the hot-water pipes keep the water at the same temperature as that of the house. If there is not room for the slate cistern on the pipes, have one sunk in the middle of the house. Cement tanks sunk in the house also answer the same purpose.

Treatment of Plants in Flower, and the best Mode of protracting their Bloom.

There are many Orchids that may be removed when in flower to a much cooler house than that in which they are grown, or even to a warm sitting-room. The following are among the advantages of keeping them during their period of flowering in a cool and dry atmosphere, rather than, as is frequently the case, in a hot and moist house : in the latter, the flowers do not last nearly so long as they do when moved to a cooler house or a warm room. Perhaps there are not many cultivators who have studied this point more than myself, and I never found the plants injured by this treatment. Some imagine, that if they are put in a cool place they will be injured ; but this, in my experience, has not occurred. During the time they are in a room, the temperature should not be below 50° at night; the room should be kept quite dry, and before they are removed from the stove they should be put at the coolest end of it; or if there be two houses, those that are in the hottest should be moved to the coolest for a few days before being taken into the room, and they should be allowed to get nearly dry, and should receive but very little water—only enough to keep the roots moist. The flowers should not receive any moisture.

The following are a few that I have tried in a room during the months of May, June, July, and August. I have kept *Saccolabium guttatum* in this way five weeks, *Aerides affine* the same time; *A. odoratum, A. roseum,* and some of the *Dendrobiums,* viz. *nobile* and *cœrulescens,* I have kept in a room four and five weeks. *D. moniliforme, D. macrophyllum, D. pulchellum, D. Ruckerii,* and *D. secundum,* last a much longer time in bloom if they are kept cool. *Brassias* and *Oncidiums, Epidendrums, Odontoglossums, Cyrtochilums, Trichopilia tortilis, Lycaste Skinneri, L. aromatica, L. cruenta, Maxillaria tenuifolia, Aspasia lunata,* and all the *Cattleyas,* succeed well in a cool room or house, and last for a much longer time in flower. I have kept *Lælia majalis* in a cool room four and five weeks, and *L. flava* will keep a much longer time in blossom than in the warm house. When the bloom begins to fade, the plants should be removed to the stove, where they may be placed in the coolest end, with plenty of shade: they ought to be kept there for about ten days, for if they are exposed to the sun they are very apt to become scorched.

Treatment of Plants previously to being taken to a Flower Show.

It is my practice to move them to a cool dry house or room for a few days. If the plants are growing in the hottest house, I move them to the cooler one. They should not receive any water for two or three days, and should be allowed just enough to keep them slightly moist. When it is probable that the plants will come into blossom earlier than is wished, the time of flowering may be successfully retarded by taking them to a cooler part of the house, or even putting them in a greenhouse, keeping them slightly shaded during the brightest part of the day. Dendrobiums are very easy to retard, if they are wanted to bloom later in the season. *Dendrobium nobile, D. pulchellum, D. macrophyllum, D. densiflorum, D. Farmeri, D. Pierardi latifolium*, these generally bloom during winter, but I have kept them back until June; and by having a succession of plants you may have the Orchid-house gay with Dendrobiums from January to June. All the Dendrobiums will bear cool treatment while at rest, and all can be kept for late flowering. The treatment they require in a warm greenhouse is to have but little water, only enough to keep them from shrivelling; the temperature should not go below 40°, and the pseudo-bulbs must be kept dry, or the flower-buds are apt to rot. When the plants are wanted to flower, move them into the Orchid-house, and keep them shaded from the sun. *Phajus Wallichii* and *grandifolius* may be kept back in the same way as the Dendrobiums.

Remarks on preparing Orchids for travelling to a Flower Show.

These plants require great care in packing and tying, for many are very tender. Their flowers being often large and waxy, some of them require more packing than others. I have seen many a plant spoiled by not being carefully packed; and it is a vexatious thing to have a fine specimen spoiled during its journey to the place of exhibition. They will travel as well forty miles as ten, if they are properly packed. I have had a good deal of this work to do, and a few hints on the subject to young beginners may be of use to them. Some kinds bear removing much better than others. *Phajus Wallichi* and *grandifolius* are both bad plants for travelling, if not well packed. I have seen fine plants brought to different flower-shows, with their flowers completely spoiled for want of careful packing, though I have shown *P. Wallichi* several years at the Chiswick and Regent's Park exhibitions, and they have always been conveyed without injury. In preparing them, I first get some strong sticks, and put one to each flower-spike; the sticks should be long enough to go into the earth, so as to remain firm; they should be placed at the back part of the flowers, and stand one or two inches above the flower-spike; then get some wadding and tie up the stick, afterwards tie the flower-spike firmly to the stick, add more wadding, putting it round every flower-stalk, and tie them firm to the flower-spike: begin at the top of the spike, and tie every flower separately, so that the flowers do not touch one another. The leaves must not be allowed to rub against the flowers. On arriving at the end of their journey, untie them, remove the wadding, and tie them in the proper form. In tying, care should be taken not to rub the flowers.

Saccolabiums and *Aerides* do not require so much packing. It is sufficient to put two or three sticks to each spike, one at each end, and one in the centre, if the spike be long; but if short, two will be sufficient. The stick, which should only be long enough to support the spikes in the drooping way in which they grow, should be fixed firmly in the basket or pot; a small piece of wadding should be placed on the top of the stick, to which the spike should be firmly tied. This will be sufficient to ensure safety. The wadding should not be allowed to touch the top part of the flowers, as it will stick to them, and be very hard to remove.

Vandas require more packing, their flowers being larger, and further apart; place some wadding between each flower on the spike, then fix some sticks firmly in the basket or pot, and tie the spike to them, without letting the stick touch the flowers : wadding should then be put in between the flowers to keep them apart.

Phalænopsis grandiflora and *amabilis* travel badly, and require much care. The best way is to get a box and set the plant in the bottom of it, which must be long enough to allow the flower-spike to lie at full length; wadding should then be placed underneath the flowers, which should lie flat on the wadding; another sheet of wadding should then be placed on the top of the flowers, in order to make them lie firm. I have also taken them to shows treated in the same way as recommended for Vandas.

Dendrobiums.—Some of these only require a stick to each pseudo-bulb fixed firmly in the pot, to tie the bulb to : such as *D. nobile, D. macrophyllum, D. Devonianum, D. moniliforme*, and sorts with similar flowers. Those varieties that flower with pendulous racemes, such as *D. densiflorum, D. Farmeri*, and others growing in the same way, require three sticks,—one to the bulb, to which the latter should be firmly tied; then put the other two to the flower spike, one at each end, in the same way as with the *Saccolabiums*.

Calanthes are bad travellers, especially *C. veratrifolia*, the delicate white soon gets injured if allowed to rub against each other ; put a stick to each flower-spike to prevent the flowers from injuring each other.

Cattleyas require to be packed very carefully ; their flowers should be tied so that they do not touch one another. I always put a stick to each flowering bulb, and tie it firmly, and a stick to each flower-stalk, just below the flower, with a piece of wadding round the stalk ; afterwards tie the stalks to the stick; neither the stick nor the leaves should be allowed to touch the flowers, or they will bruise.

Oncidiums travel well ; they only require a strong stick

to each flower-spike, with a piece of wadding round the stick when tied.

Sobralia macrantha is a bad plant to travel, if not properly tied. There should be a strong stick to each flowering bulb, and tied firmly; and also one to the flower-stalk, with a piece of wadding close to the flowers; then tie the stalk firmly to the stick, and allow nothing to rub against the flowers.

Peristeria elata.—This should be treated in the same way as *Phajus.*

Cypripediums require a small stick to each flower-stalk. All the Lycaste, and other Orchids that flower in the same way, require similar support for their flowers.

The best mode of conveyance for Orchids is a spring van with a cover on the top. In placing the plants in the van, I always put some hay between each pot, to keep them firm and prevent their rubbing against each other.

Treatment during the Time of Rest.

Rest, as has been elsewhere stated, is of great importance to Orchids, as well as other things. No plant will continue long in good health without it. My practice is to give them a long season of rest, generally from November to the middle of February. During this time the temperature of the East India house should be regulated so as to keep it as near as possible at 60° by night, and 65° by day; but by sun heat the temperature will rise a few degrees higher. Air must then be given so as to keep it about 65°; but a few degrees of solar warmth above this point will do no harm. A little air should be given on every fine day, in order to dry up damp; but the air must be admitted close to the hot-water pipes, so that it may become warm on entering the house. As to those plants that come from the hotter parts of India, the temperature should not be allowed to go below 60°. The Mexican house should range from 50°

to 55° by night, and from 55° to 60° by day; this should not be allowed to go below 50° at night.

Rest is induced by lowering the temperature, and withholding water: during this season plants should only receive sufficient water to keep them from shrivelling. There are, however, some that will grow during the winter months, as many of the Aerides, Vandas, Saccolabiums, Phalænopsis, Zygopetalums, and similar kinds. These will require water at the roots to keep them increasing, but care should be taken not to wet the young shoots, for if they get wet they are very apt to rot at this season of the year. Those that are growing should be placed at the warmest end of the house.

Some Orchids are deciduous, losing their leaves after they have finished their growth. To this class belong *Cyrtopodiums, Barkerias, Cycnoches, Phajus albus*, some of the *Dendrobiums, Pleione maculata, Wallichiana*, and many others. I always place these so that they may have as much light and sun during their season of rest as possible. This is the only way to ripen their pseudo-bulbs, which causes them to grow stronger and flower more freely. These plants require but very little water when at rest. But when such plants as Vandas, Angræcums, Aerides, Saccolabiums, and Phalænopsis are at rest, they should never be allowed to get too dry at the roots: the moss should always be kept a little damp; for the stems and leaves are very apt to shrivel if kept too dry, and this often causes them to lose their bottom leaves; and they require but a short season of rest. Those which are growing on blocks will require more water than those which are in pots or baskets, and they should be watered about twice or three times a-week if the weather be fine, but in dull weather they will not require it so often. The water should be poured over the paths and walks every fine morning, with a view to create a moist atmosphere, but the moisture in the house must be regulated according to the weather outside;—this is an important point to be attended to : if the weather be dull outside, the house inside must be kept dry. The glass and roof should always be perfectly clean, so that the plants may have plenty of light and sun : it is my practice, as soon as the season of rest commences, to wash every piece of glass on the house, and also the woodwork.

Insects.

Orchids are liable to be injured by many sorts of insects, such as red spider, thrips, mealy-bug, white and brown scale, woodlice, cockroaches, and a small shell snail. Cockroaches are among the greatest plagues we have to deal with; they will do a great deal of mischief in a few nights, if they are not sought after, on every opportunity. The food they like best is the young tender roots and flower stems. I have seen the roots of a plant completely eaten off in one night. The only way to keep these insects under is by constantly looking after them, both by night and day. Search for them in the evening by candle-light, and in the daytime by moving the pots and baskets under which they harbour. They leave their hiding-places in the evening, to seek after food, and it is then that they are most easily caught. Chase's Beetle Poison, a mixture sold in boxes, is a capital thing to destroy them, if laid in different parts of the house in the evening; lay it on some pieces of tile or slate two or three nights a week; then move it, for a week, and keep on every other week until you find that they are destroyed. By using this at times they may be kept under : collect the pieces employed every morning, and put them down again in the evening. It is also a good plan to lay some damp moss in the hottest part of the house; I have killed many in this way, looking the moss over every two or three days. I have also destroyed them with a mixture of honey, lard, and arsenic, the latter in very small quantity.

Place some of this in oyster shells in different parts of the house. Some growers mix the arsenic with tallow put on a stick, which is stuck in the pots: care, however, must be taken that the mixture does not touch the leaves or bulbs of the plants.

Small ants are another pest in the Orchid-house, as they carry the dirt to the flowers, spoiling their appearance. The best thing I know of to catch these little and troublesome insects is to cut apples in halves, scoop out some of the inside, and lay the pieces in different parts of the house, looking them over very often. I have destroyed hundreds this way in a very short time. Treacle is also a good thing to trap these pests; place some in a bell glass where they frequent, they are fond of anything sweet; they go to feed, get into the mixture, and cannot get out again; it sticks to them and thus causes death.

The woodlouse and the small shell snail are also very destructive. These, like the cockroach, are very fond of the young roots; they may be trapped by cutting some potatoes in two, scooping out the inside, and placing them on the pots and baskets, looking over them every night and morning till you clear your house of these vermin. Turnips cut in slices will answer the same end. Toads are very useful in catching insects; a few of them in a house do good.

The best way of getting rid of red spider and thrips is by constantly washing the leaves with clean water, and by fumigating the house with tobacco. My method is to fill the house with tobacco-smoke three or four times; this should be done every two or three days till the insects are quite destroyed. Evening is the best time to do this. It is also a good plan to mix some lime and sulphur together, and rub it on the pipes in different parts of the house, taking care not to use too much; and it should be used only when the pipes are warm. There should be a good supply of moisture at the same time, but not too much heat. The green fly, which makes its appearance in spring on the young flower-buds may also be destroyed by tobacco-smoke. I always use tobacco-paper, which I keep ready prepared for use.

The brown and white scale and mealy-bug I keep under by constantly sponging the leaves and bulbs with water; the white scale is very troublesome if not looked after. I have found a little soft soap a good thing to destroy white scale; mix a little with water, and rub it over the leaves and bulbs; let it remain on for a day, then wash it off, and all will be destroyed: care should, however, be taken not to use it too strong. Cattleyas are very subject to this; the following is a recipe for destroying these insects :—viz. to one gallon of rain water add eight ounces of soft soap, one ounce of tobacco, and three table-spoonfuls of turpentine; stir well together, and leave the mixture for forty-eight hours; then strain it through a cloth: what you have to spare must be bottled off. It is necessary to rub the plants over two or three times, if they are much infested, but once will be sufficient in most cases.

This recipe is also a cure for thrips. I have been informed by Mr. Smith, gardener to Sir James Watts, of Abney Hall, near Manchester, one of our best plant growers, that he has used this mixture ever since I wrote the first edition of this book for destroying thrips on his Azaleas; he gets a large tub and mixes the ingredients as recommended, he then dips his Azaleas in it, completely destroying this pest; and he also uses it for his stove-plants, large specimens of which are cleared of mealy-bug and scale by the use of this mixture. I am therefore glad to find that others have found it useful as well as Orchid-growers. It will prove a great saving for those who grow large plants for exhibition, or even those who have a quantity of small plants, to clear of these pests without much labour. We have kept mealy-bug away from this place by constant watching; and when we buy a plant that is infested with it, we take care to clean it well before placing it in the stove or Orchid-house.

Rot in Orchids.

Orchids are subject to disease in their leaves and bulbs, especially during the damp months of winter. The rot, which is apt to assail the thick, fleshy bulbs, is caused by too much moisture in the house; when the heat is low, sometimes the drip from the glass will fall on the crowns of the pseudo-bulbs, and they soon rot. Steam is also very bad in a house during the winter, particularly to such plants as Cattleyas, Peristerias, Odontoglossums, and any other Orchids that have fleshy bulbs. When rot attacks the bulbs it should be seen to at once, and may be easily stopped by cutting the diseased part entirely away with a sharp knife. No portion of the diseased or decayed bulb should be left; the wound should then be filled up with sulphur, keeping it dry. When the leaves begin to rot, the diseased part should be cut clean away, and a little sulphur rubbed on the part that is cut, but not in such a manner as to let the sulphur get to the roots of the plants.

When any part of the fleshy pseudo-bulb of the above-named or similarly-formed bulbs becomes discoloured, and the dark or discoloured part appears moist or wet, especially if any fluid exudes from it on pressure, the wet or discoloured part should be immediately cut out, or there is danger that the bulb will be destroyed, as the rot is often much more extensive within the bulb than the discoloured appearances on the outside would seem to indicate. The plant should also be removed to a drier and cooler place, and water given with the utmost care.

Spot in Orchids.

A great deal has been said and written respecting this disease which I have seen in several places, and in other plants as well as in Orchids. I have not experienced much of it myself, but I have given advice respecting its cure, and have found in most cases the plants to outgrow it.

In 1860, a gentleman bought some Phalænopsis of me, which were the finest grown plants I had seen, and they did beautifully with him for some time, growing very fast, in fact too fast, for they got sappy, and their leaves became spotted, as did also some others he had. I went to see them, and he asked my opinion respecting them. Upon inquiring how they had been treated, he told me that he gave them a great deal of water over the leaves, and kept the roots wet; this was during winter. I went several times to see them the following spring, and advised him to treat them as recommended for Phalænopsis. (See p. 132.) These plants have quite outgrown the spot, and are now among the finest round London.

When at Hoddesdon, I had two Phalænopsis which went in the same way; they got spotted in the winter. I cut off one of the leaves, and sent it to Dr. Lindley for his opinion as to the cause of the evil. His reply was, that the plants had been kept too moist during the cold dark days of winter, —a fact which I have never lost sight of. He was quite right, for it was a sharp winter, and I had kept these two plants wet, by placing the bottom of the block on which they grew in a pan of water to keep off cockroaches. In that way too much moisture crept to the roots, and, being in the winter time, doubtless caused the leaves to become spotted.

However, by following the treatment recommended in my remark on Phalænopsis, the plants soon recovered; but if steps of that kind had not been taken in time the disease would have gone too far, and probably killed the plants. Much injury is done by keeping the plants too wet at the roots in dull weather; in fact, too much moisture in that way is injurious at any time, especially to those that grow on branches of trees, when they come to be confined in pots and baskets with a host of wet material about them, which is unnatural. My practice is to give but little water at the roots during winter, and not so much as many people in summer, because I have seen the ill effects of it. The treatment I recommended some years ago I still follow, and with uniform success.

Some Orchid growers give more heat and moisture than I recommend, and the plants have done well for a time; but under such exciting management they are apt to become spotted, and get into a permanently bad state of health; in short, the least chill, after so high a temperature, is liable to induce disease. The great secret in the cultivation of these, as well as all other plants, is a proper house, sufficient ventilation, heat and moisture, and good glass, without drip. Let the temperature throughout the whole year be in accordance with directions laid down in this book, which is the result of long experience and close observation.

Propagation.

There are different modes of propagating the various kinds of Orchids; some are easily increased by dividing them into pieces, or by cutting the old pseudo-bulbs from the plants after the latter have done blooming: such plants as Dendrobiums are increased in this way. The best time for dividing the plants is just as they begin to grow, or when they are at rest; they should be cut through with a sharp knife between the pseudo-bulbs, being careful not to harm the roots: each piece should have some roots attached to it. After they are cut through they should be parted, potted, and put into some shady part of the house, without receiving much water at the roots till they have begun to grow and make fresh ones, then they may have a good supply. *Dendrobium nobile, Pierardi, pulchellum, macrophyllum, Devonianum,* and similar growing sorts, are easily propagated. This is effected by bending the old pseudo-bulbs round the basket or pots in which they are growing, or by cutting the old flowering bulbs away from the plant, and laying them on some damp moss, in a shady and warm part of the house, with a good supply of moisture. After they break and make roots they may be potted or put in baskets. Such as *D. Jenkinsii, D. aggregatum, D. formosum, D. speciosum, D. densiflorum*, and similar growing sorts, are increased by dividing the plants.

Aerides, Vandas, Angræcums, Saccolabiums, Camarotis, Renantheras, and similar growing kinds, are all propagated by cutting the tops off the plant just below the first root, or by taking the young growths from the bottom of the plant. After they have formed roots, they should be cut off with a sharp knife, and afterwards put on blocks or in baskets with some sphagnum moss, and kept in a warm and damp part of the house, without receiving much water till they have begun to grow, when they may have a good supply. Odontoglossums, Oncidiums, Zygopetalums, Sobralias, Trichopilias, Stanhopeas, Schomburgkias, Peristerias, Mormodes, Miltonias, Lycastes, Leptotes, Lælias, Galeandras, Epidendrums, Cyrtopodiums, Cyrtochilums, Cymbidiums, Cycnoches, Coryanthes, Cœlogyne, Cattleyas, Calanthes, Brassias, Bletias, Barkerias, are all propagated by dividing them into pieces, each having a portion of the roots attached to it, and a young bulb on the pseudobulb.

Phajus albus is very easily increased. The best way is to cut the old pseudo-bulbs off after the young ones have begun to flower, that is, just before the plant has made its growth. The pseudo-bulbs should be cut into pieces about six inches long, and then put into a pot in some silver sand, with a bell-glass over them till they have struck root; they should then be potted in some fibrous peat, and should have good drainage, and a good supply of water in the growing season.

Some of the Epidendrums are easily propagated, such as

cinnabarinum and *crassifolium;* these will form plants on the tops of the old flower-stalks; they should be left till they have made their growth, they should then be cut off and potted, and they will soon make good plants. Some Dendrobiums will also form plants on the tops of the old pseudo-bulbs, and they should be treated in the same way.

Mode of producing Back Breaks.

There are many of our Orchids that will keep on growing year after year, and yet produce only one flowering bulb each year; but if the plants are cut they will produce back breaks, increasing, and soon make fine specimens. This is the way to produce such plants as are seen every year at the London Exhibitions. Some plants are more easy to increase than others. The Cattleyas are of this kind. When you have a plant that has back bulbs, if there are about four, cut the plant in two between the bulbs, but not to disturb the plant; let the bulbs keep in the same place. The best time to cut all Orchids is during their season of rest, or just as they are beginning to grow. All other Orchids that have bulbs should be treated in the same way, if it be desirable to increase them.

On the Mode of making Baskets, and the best Wood for that Purpose.

Blocks or baskets are more natural for true air-plants, such as *Vandas, Saccolabiums, Aerides, Angræcums, Phalænopsis, &c.;* when planted in baskets or on blocks, they send out their roots much stronger into the air, and suck up the moisture, whereas, if their roots are covered too much, they are very apt to rot. Various materials are used for forming baskets; some are made of copper wire, which is very durable: but I prefer those made of wood, on account of their rustic appearance, and the roots like to cling to the wood. The best kind of wood is maple or hazel, and the best baskets are those of a square shape. The wood should be cut into the lengths as the size of the basket may require; but do not make them too large : there are two objections to this—one is, that they take up much space; the other, that the plants do not require much room. After the wood is cut into proper lengths, the pieces should be bored within one inch from the ends, taking care to have all the holes bored the same distance : there should be four lengths of copper wire, one for each corner; the wire should be put through each piece of wood, and brought up to form the handle for suspending the plants from the roofing. Iron wire should never be used in making baskets, for it is probably injurious to the plants.

The best kinds of wood for blocks are Acacia, Apple, Pear, Plum, or Cork, if it can be obtained. The wood should be cut into lengths, suitable for the size of the plants; get some nails, and drive one at each end with some copper wire to form the handle, wind the wire round each nail, and leave the handle about ten inches high. Small copper nails are the best by which to fasten the plants on the blocks.

Seedling Orchids.

Few Orchids have as yet been raised from seed in this country; a large field is, however, open for all who take an interest in hybridising this singularly beautiful tribe of plants. Some time ago a gentleman said to me he should like to be in a country where the Orchids grew in a wild state, in order that he might have a chance of hybridising them; his ideas were that something really good might be obtained; and no doubt he was right, for how seldom is it that we flower two Orchids alike ? I have seen at least twenty varieties, or nearly so, of *Cattleya Mossiæ* in bloom at one time; some had white petals and rich crimson lip; others rose-coloured petals and yellow lip, all differed more or less from each other; in fact, nearly all species of Orchids have varieties. Four flowers taken from four different plants of *Phalænopsis amabilis* were brought me the other day by a gentleman, and no two of them were exactly alike. The same may be said of the new *Phalænopsis Schilleriana*. I have only seen five plants of this in bloom, and all of them differed in colour, shape of leaf, and flower; all were, however, handsome. In a wild state varieties appear to be unlimited, crossed, and recrossed, as they doubtless are, by insects. Who, for instance, would have thought, a few years ago, of receiving so splendid an importation as *Phalænopsis Schilleriana*, beautiful not only in blossom but in foliage ? I have heard that there exists a scarlet Phalænopsis ! Let us hope yet to receive it : what a contrast it would make with the white and mauve coloured kinds now in cultivation. In this country Mr. Dominy has succeeded in raising some pretty varieties of Cattleya, Calanthe, Goodyera, &c. I trust he may persevere in the good work, and produce us something new in other genera; many kinds seed freely if the flowers are set, producing many seeds in a pod. When ripe the seed should be sown ; but it requires great care, as it is not so easy to raise as that of many other kinds of plants ; some of the kinds are a long time in germinating : I have known Orchid seeds to lie twelve months before they made their appearance. To watch their progress when up is, however, highly interesting —first, the formation of pseudo-bulbs, then their advancement towards flowering, are processes full of pleasure yielding anxiety. The best place to sow is on the top of an Orchid pot, where the seeds will not get disturbed ; let the peat be in a rough state : do not cover the seed, but give a little water with a fine rosed-pot, just to settle it in the peat ; some rough blocks of wood on which another plant is growing afford a capital situation to sow upon : they should always be kept a little moist ; and of such as are sown on pots in the same way, when the plants are strong enough, pot them off into separate pots, or place them on blocks in material already recommended ; in potting and taking them up care must be taken not to break the roots ; by hybridising the finer kinds you are most certain to get fine flowers. Of Cattleyas we have only one which is not worth growing, and that is *C. Forbesii*, yet that is better than many other Orchids in cultivation; therefore let many begin to raise hybrids, not only with the view of obtaining finer flowers than we already possess, though that would be a real acquisition, but for the additional purpose of raising sorts that might succeed in cooler houses; *Odontoglossum grande* and many others, for instance, do better in a cool house than in a warm one; *Cypripedium insigne* will also thrive well in a greenhouse;—if, therefore, we could cross this with some of the other kinds, such as *C. grandiflorum*, *C. hirsutissimum*, *C. Lowii*, or *C. barbatum superbum*, something good might be the result. There is also our hardy *Cypripedium spectabile*, which might be induced to play an important part in the operation. *Phajus grandifolius* and *Wallichii* are likewise two noble plants for winter decoration which do well in a warm greenhouse. Might not these be crossed with *albus*, which is white, and something new be realised, provided they could be had in flower at the same time? Moreover Lycaste will do in a cool house, as, for instance, *L. Skinneri*, which is one of the finest. Many splendid varieties of this are now cultivated; I saw eight varieties of it the other day. This plant is reported in the "Gardeners' Chronicle" to have been in a room in flower for seven weeks—a fact which shows what might be done with these fine plants in a cool house. I have, myself, had this Lycaste all the winter in a greenhouse, where it flowered in great abundance, as many as from thirty to fifty blossoms being open at one time. I exhibited one plant of it at Regent's Park spring show last year, with as many blossoms on it as I have just mentioned, on which occasion a medal was awarded for its magnificent flowers and colour. We must be particular, however, in keeping the flowers dry when in a cool house, or else they are apt to become spotted.

Lycaste Skinneri, says the "Gardeners' Chronicle," seems about to have as great a future as the Tulip. Already something like a dozen varieties of colour are known among its exquisitely beautiful flowers, and we can entertain no doubt that it will break into plenty more, especially if recourse is had to hybridising. From deep rose to a skin only less white than the Hawthorn we have a complete set of transitions, and this is a plant conspicuous for its fine broad foliage, and most glorious in its ample floral garments.

It is not, however, wholly on account of its disposition to reward us by an endless variety of colour, and perhaps form, that we wish to attract attention to it, but because of all tropical Orchids it is one of the hardiest and most easy to cultivate. This has been very decisively shown by some late experiments by Mr. Skinner, to whose untiring energy we English owe this and many another treasure. In a note received from him the other day he writes as follows :—

"On the 2nd February, 1861, I received from Mr. Veitch a fine specimen with seven flower spikes all out, and took it to Hillingdon Cottage, placed it on the drawing-room table in an ornamental pot, and gave it every three days or so about four tablespoonfuls of water, occasionally wiping the leaves with a wet sponge when the dust got on them. There this plant stood throughout the severe weather we had—a fire in the room only during the afternoons and evenings, and on some days none at all. It never showed the least decay until the 16th of May, when it was for some purpose or other put into the greenhouse, and our gardener sprinkled water over it along with the other plants. Next morning I was shocked to see the flowers all with brown spots and withering. On the 18th May I took it back to Mr. Veitch, still in full bloom (seven spikes) to bear testimony to its condition, and it lasted, though then much injured, a week on the stand by the seed-room in their place. This experiment induced me to try again. On the 18th December,

D

1861, I brought down to this place two fine plants of the *Lycaste*, and two plants of *Barkeria Skinneri*, both in full bloom. Having been absent (with the exception of three days in January) since, I have had no control over them, but my sister followed the same plan as at Hillingdon, only with the *Barkerias*, which are attached to blocks, dipping the whole block into water for a few minutes every four or five days, according as we have much or little sun; and as the plants are now before me, I give you their condition. One of the *Barkerias* is as perfect as the day I brought it here. The other has all gone off within the last few days. One *Lycaste* is perfect, and as beautiful as the day I brought it here; the other has *lost* one flower, I fear, by some accident, the other flower still good, but evidently a little 'shady;' this plant has two flower stems coming on, and will bloom in a fortnight if we pushed them by more moisture. I expect frost has got on it after watering, for it stands close to the window in the drawing-room, and this room, though smaller, is similarly treated to the one at Hillingdon—fires in the afternoon and evenings, with a southern aspect. What a treat to me is this, and I think you should know it, for people have said—'I love Orchids, but hate the stew-pans one has to view them in.'"

It is clear that for *Lycastes* and *Barkerias* "stew-pans" may be dispensed with. Plenty of Orchids like these are to be found in our gardens, brought from the Nubes or Highlands of Mexico and from Central America. It is also probable that mountain species of India, such as the delicious *Cœlogynes*, will thrive under the same treatment, and, if so, one more class of enjoyments is provided for the lovers of flowers.

Surely this is news worth telegraphing through the whole horticultural world! What a charm for a sick room! What a pet for the poor invalid who has nothing to love except her flowers! Imagine the pleasure of watching the buds as they form, visibly enlarging from day to day, and the slow unfolding of the perfect blossom, and then the delight at seeing it some morning, stimulated by even a winter's sun, suddenly throwing back its green cloak and displaying the wondrous beauty of its richly tinted lining. It is almost worth being ill to enjoy such a scene.

Encouragement like this must surely add new life and vigour to Orchid growing. I hope, therefore, that many may be induced to try their skill; the great secret is robust growth in summer, when there is plenty of heat in the greenhouses; but for further instruction, see remarks on the cultivation of the *Lycaste* (p. 116), and also on the treatment of plants in flower.

ORCHIDS AT PRESENT IN CULTIVATION.

In the following brief descriptions of all the best Orchidaceous plants I have seen, a general account is given of the distinctive features of each genus, which is followed by a more particular description of each species, together with an account of the mode of treatment which, after considerable experience, has been found best adapted to each individual plant.

ACINETA.

To this singular genus belongs several species, but only three that I have seen are worth growing. They produce their flowers from the bottom of the basket. They are all evergreen, with short pseudo-bulbs, and leaves about a foot high; they are of easy culture, and are best grown in baskets with moss and peat. They require a liberal supply of water at the roots during their period of growth, afterwards less will suffice; they will do in either house suspended from the roof, and all of them are propagated by division.

A. Barkerii.—A curious Orchid from Mexico, producing

from the bottom of the basket spikes of yellow flowers, each about a foot in length, which, if kept dry, will last a long time in perfection. It blooms during the summer.

A. densa.—A beautiful and distinct species, also from Mexico, with yellow and crimson flowers, which are produced from the bottom of the basket, and, if kept dry, the flowers will last a long time in perfection.

A. Humboldti.—A strong-growing species from La Guayra. Flowers in the same way as *Barkerii*, but about a month earlier; the long spikes of flowers are of a deep chocolate colour, spotted with crimson. It lasts only a short time in perfection.

In order that the flowers in this class as in all others may be preserved in bloom, care should be taken not to wet them while watering the plants.

AERIDES.

Aerides are among the most beautiful of Orchids, many of them uniting every good quality that a plant can possess,—rich, evergreen, and regularly-curved foliage,—a graceful habit,—flowers of peculiar elegance. Even when not in bloom the plants themselves are interesting objects, and give a sort of tropical character to the collections in which they are found. The stem of the plant is straight or slightly bent, with leaves attached on opposite sides, and the plant is nourished by large fleshy roots, shooting out horizontally from the lower part of the stem. The flowers, which are rich and waxy, proceed from the axils of the leaves, and extend in delicate racemes one or two feet in length, while their fragrance is so abundant as to fill the house in which they grow with grateful perfume.

These plants are of easy culture, and if properly attended to are seldom out of order. They are found in the hottest parts of India and other warm countries, growing on the branches of trees, generally on such as overhang streams of water; and to grow them in anything like perfection, the climate in which they grow wild must be imitated as nearly as possible. I find they succeed best with a good supply of heat and moisture during their growing season, which is from about March till the latter end of October. During that time I keep the temperature, by day, from 70° to 75°. It may be allowed to rise to 80° and 85°, or even higher will not do any harm, provided the house is shaded from the rays of the sun. The night temperature should range from 65° to 70° in March and April, and afterwards it may rise five degrees higher.

Some grow their *Aerides* in baskets made of wood, but they may be grown in pots,—a mode of culture successfully followed by many of our Orchid growers, especially those who exhibit; for, when in pots, the plants are more easily moved about. Another reason is, baskets soon decay; but if the plants are not required for exhibition, I should advise some of the small growing kinds to be grown in baskets, in which they have a fine appearance, especially if there is room for suspending them, letting the roots grow out of the basket. All the kinds will succeed well in this way. In pots give good drainage—about three parts full of small potsherds mixed with moss—and when the moss begins to decay fresh should be given to keep all healthy. They all do well on blocks of wood, but to grow them in perfection they require care as regards moisture. In their native country they are found growing on branches of trees. Sphagnum moss and broken potsherds have proved the best materials for filling baskets. They require frequent watering at the roots during the growing season; indeed they should never be allowed to become dry, not even during their season of rest, as they are liable to shrivel and lose their bottom leaves. *Aerides* require but little repose, and the moss should always be kept damp; but during the dull months of winter no water should be allowed to lodge on the leaves or heart of the plant, as it would be very apt to rot them. The plants, if not in pots, should be suspended from the roof, but not very near the glass, lest they should be affected by cold; and they should be kept perfectly free from insects, especially the different kinds of scale. There is a small kind which is apt to infest them, and which, if allowed to get ahead, will make the plants look yellow and unhealthy. It may be kept under by constantly washing with water and a sponge. These plants are propagated by cutting them into pieces, with roots attached to each piece. Some kinds, however, are shy in throwing up young shoots, and this makes these sorts very scarce. The *A. odoratum* division is the most easy to increase, and *A. crispum* sends out roots more freely than some others. If the plants ever get into an unhealthy condition, the best way is to cut the bottom off the plant, and give fresh moss, with plenty of water at the roots.

A. affine.—A handsome free-flowering Orchid from India, with light green foliage a foot long, and pink and white flowers, produced on long branching spikes in great profusion. I have seen spikes of this two feet long, and three and four branches on each spike. This has been exhibited with from thirty to forty spikes. It grows from two to three feet high, and, if true, makes one of the finest plants for exhibition, continuing in blossom three or four weeks.

A. affine superbum.—A splendid variety of the former, the colour of which is richer, and the flowers much larger, and more compact in growth than *affine;* a free-flowering branching variety.

A. crispum.—A truly beautiful free-growing Orchid from India, with purple-coloured stem, dark green foliage, ten inches long; the blossoms, which are abundant, are white tipped with pink. Flowers in June or July, and lasts two or three weeks in good condition. The spikes of flowers are long, and very distinct from any of the other kinds.

A. crispum, var. Lindleyanum.—A charming kind, with

AERIDES.

blooms of a fine rich colour; growth similar to that of *crispum*, and it blooms about the same time.

A. crispum pallidum.—A variety of *A. crispum*, grows to about the same height, and flowers at the same time. The blossoms are of a lighter colour.

A. crispum, var. Warneri.—A splendid free-flowering variety from India. In leaves and stem it closely resembles *A. crispum*, except that the leaves are smaller and more slender in growth; the blossoms, which are produced in June and July, are white and rich rose colour, and they last three or four weeks in perfection.

A. cylindricum.—A very distinct growing plant, having a habit like that of *Vanda teres*, but not so strong; the flowers, which are produced in pairs from the axils of the leaves, are as large as those of *A. crispum*, and are of a white and pink colour.

A. Fieldingi.—A magnificent free-flowering Indian species, of which there are many varieties both in growth and flowers; it grows from two to three feet high; some of the varieties have dark green foliage, while others are of a lighter shade; the leaves, which are broad, are eight or ten inches long; the spikes sometimes attain a length of three feet, and are branched; the flowers are white and rose colour, unusually large, and are produced during May, June, and July, continuing in bloom three or four weeks. This makes a fine exhibition plant.

A. Larpentœ.—A fine Indian free-growing plant, with dark green leaves, ten inches long. The flowers are numerous on a single spike, of a cream and light rose colour. It blooms in June, and lasts two weeks in perfection. This is a distinct plant, and was first flowered by Mr. Eyles, then gardener to Lady Larpent, and shown at the Regent's Park Exhibition in 1847, when it received the first prize as a new plant.

A. Lobbii.—A free handsome flowering species from India, producing long spikes of the same coloured blossoms as *affine;* foliage light green, about eight inches long. Altogether a very showy and scarce kind, of which there are several varieties. Slow growing, but very compact in habit.

A. maculosum.—A lovely dwarf Orchid from Bombay, with dark green leaves, eight inches long, close and compact, stiff growing, with light-coloured flowers, spotted all over with purple, and a large purple blotch on the lip. Blooms in June and July, and, if the flowers are kept dry, lasts four weeks in perfection. The colour of the bloom is very striking.

A. maculosum, var. Schröderi.—A magnificent free-growing plant from the hills near Bombay, much stronger than *A. maculosum*, and more in the way of *A. crispum*, with dark green foliage, ten inches long; the flowers are very delicate, the sepals and petals almost alike; white, tinged with lilac and spotted with rose; the labellum a beautiful rose colour. It flowers in June or July, lasting three weeks in perfection. This was first flowered by Mr. Plant, then gardener to J. H. Schröder, Esq., of Stratford. It is supposed that there was only one plant imported. The stock at present in this country is from the one plant.

A. McMorlandi.—A magnificent species from India, compact in growth, and having bright green foliage, about ten inches in length; blossoms freely, producing long branching spikes of peach and white flowers, in June and July, and continues three or four weeks in perfection. The only plant I know of this is in the collection of E. McMorland, Esq., Haverstock Hill, in compliment to whom it is named.

A. nobile.—A magnificent free-flowering species from India, in the way of *suavissimum*, but with flowers larger, and of a better colour, and in growth it is much stronger. I have seen spikes of this from two to three feet long. Blooms in June, July, and August, and keeps in perfection three or four weeks.

A. odoratum.—A good, old, free-growing species from India, one of the most abundantly flowering of this genus, having pale green foliage, blooming in June or July, and remaining two weeks in good condition. The blossoms are white stained with pink. I have seen specimens five feet high and four feet in diameter, which produced thirty or forty spikes of bloom every year; altogether a noble plant.

A. odoratum cornutum.—A handsome free-flowering variety from India; in growth distinct from the former; spikes about twelve inches long, furnished with pink and white coloured flowers, which are produced in May, June, and July, and continue upwards of three weeks in bloom.

A. odoratum, var. majus. — Like *odoratum* in growth, and differing only in the larger size and longer spikes of flowers. There is another variety of *odoratum*, called *purpurescens*, the flowers of which are of a much darker pink colour and the leaves broader. This is a desirable plant, and is rather scarce.

A. quinquevulnerum.—A splendid free-flowering Orchid from Manilla, with light green foliage, about one foot long: less compact than many other species. Sepals and petals white, spotted with purple; the top of the lip is green, the sides pink, and the middle a deep crimson: it blooms in July or August, and lasts two or three weeks in bloom. There are two varieties; one with much lighter coloured flowers than the other.

A. quinquevulnerum album.—A white variety of the preceding, producing long spikes of white flowers; grows like *quinquevulnerum*, and blooms about the same time. The first time this plant came under my notice was at Nonsuch Park, Cheam, the seat of W. F. G. Farmer, Esq. It is by no means plentiful.

A. roseum.—A beautiful dwarf-plant from India; leaves a foot long, spotted with brown. A slow-growing species, bearing light rose-colour flowers, spotted with darker spots of rose, in June and July. This plant, which does not root freely, requires less moisture than any of the other kinds.

A. roseum superbum.—A fine variety, much stronger in growth than *roseum;* flowers also larger, and of a richer colour. The spikes of this, as well as of *roseum*, are apt to damp off at the ends before the flowers open, which is often caused by too much moisture having been given. It continues a long time in perfection.

A. suavissimum.—A distinct and desirable species, of free growth, with light green foliage spotted with small brown spots, ten inches long. The sepals and petals are white, and the lip has a blotch of yellow in the centre edged with white. Blooms in July, August and September, and lasts in good condition three weeks. Flowered two varieties of this species, one of which was much better than the other, having pink spots on the end of each petal, which makes a more showy flower.

A. Veitchii.—A charming species from India, with leaves about eight inches long, dark green and covered with small spots; flowers of a white and pink colour like so many small shells, so beautiful are they in appearance. Blooms during June and July, and lasts about three weeks in good condition.

A. virens.—A lovely plant from Java, with light green foliage, eight inches long; the flowers are of a light peach colour, spotted with purple; the lip is spotted with crimson. Blooms in May and June; the flowers remain long in perfection. A desirable species.

A. virens grandiflorum.—A magnificent variety from India, whose flowers are larger than those of *virens*, of a white and pink colour, and more graceful in their manner of growth. The only plant I have seen of this is in the collection of J. A. Turner, Esq., Pendlebury, near Manchester, where it blooms during April and May, and continues from three to four weeks in beauty.

A. virens superbum.—This is another fine variety from India, the growth of which is the same as that of *virens*, except that the spikes and flowers are longer and of a brighter colour. Blooms about the same time as *virens*, and remains long in perfection.

A. Williamsii.—A distinct and charming kind from India, with broad dark green drooping foliage. The spikes of flower are produced in great abundance, measuring from two to three feet in length, and branched; colour pinkish white; very scarce. I am only acquainted with one plant, which is in the collection of C. B. Warner, Esq., at Stratford, where it blooms in June and July, and makes a fine Orchid for purposes of exhibition, on account of its free-flowering; its delicate colour also associates well with that of other kinds.

AGANISIA.

A. pulchella.—A pretty dwarf Orchid from Demerara, eight inches high, the only species of this genus that I know; it produces from the bottom of the bulb a spike of flowers, which are white, with a blotch of yellow in the centre of the lip. It blossoms at different times of the year, lasts two or three weeks in perfection, and is best grown in a pot, with peat and good drainage. It requires a liberal supply of water at the roots, and the hottest house. It is a very scarce plant, and is propagated by dividing the bulbs.

ANÆCTOCHILI OR VARIEGATED ORCHIDS.

The following short and plain description of all the *Anæctochili* I have seen grown may, I hope, prove serviceable to those who may be beginning their cultivation, as well as to others who, having made a commencement, have not completed their collections; I have also added the cultivation which, from long experience, has been found to suit their wants. Many are, however, not yet in cultivation, and there are others in catalogues that I have not seen.

Among Orchids, as well as among most other orders of the vegetable kingdom, there are variegated varieties, and these, like most other variegated plants, generally bear flowers small and unattractive compared with the beauty of their foliage. To this rule the charming *Phalænopsis Schilleriana*, described by me at p. 137, is, however, an exception; some of the *Cypripediums*, too, have both fine foliage and flowers. The genus *Anœctochilus* is one of the most remarkable of this handsome class of Orchids, and to its cultivation, which is not generally well understood, I will now address myself. All the varieties are remarkable for compact dwarf habit, perfect form, and great beauty: they vary in height from two to six inches, and their leaves, which are well defined, vary from two to five inches in length, including the stalks, which, like the stems of the plant, are short and fleshy. The foliage of all the species is singularly beautiful; in some of the varieties it resembles the richest olive or rather purple velvet, regularly interspersed with a net-work of gold. In others the leaf is of the most lively green, covered with silver tracery. As regards cultivation, the plants require sand and peat mixed with moss; the white ground colour from which they spring enlivened occasionally by small growths of moss, sets off the plants to much advantage, especially when looked at through bell-glasses, under which the delicacy, richness, and softness of their appearance are increased. Few visitors walk through a house containing any of these plants without bestowing on them more than ordinary attention, and expressing admiration of their elegance, richness, and beauty. All of them demand treatment very different from that of any other Orchid; different growers operate in different ways, but I have not found any mode of management to succeed better than the one I first laid down about twelve years ago, and which I know is being followed by many who have *Anœctochili* growing in great perfection. They are difficult to cultivate, and many fail with them, a circumstance I attribute to keeping them too close. The case in which they are grown should always have a little air by tilting the glass about one or two inches; this will benefit them very much and make them grow stronger: when too much confined in the case or bell-glass they grow up spindling and damp off in the stem; the latter being fleshy requires more substance and hardihood. I have seen *Anœctochili* grown in bottom heat, which I find to be injurious: they succeed in it for a time, but not long; they grow too fast, and become so weak as to often die altogether. What I had in bottom heat I removed to a cooler house without it, and I found that they improved very fast under cool treatment.

The finest collection I ever saw belongs to J. A. Turner, Esq., of Manchester, and is under the care of Mr. Toll, his gardener; the plants are grown under bell-glasses in a little top heat but no bottom heat. I sold Mr. Turner a plant of each kind about three years ago; they have been grown and propagated, and now he has large pots full of them, many plants of a kind being placed together. This shows how well they may be grown without bottom heat. I also have seen them doing well in other places without bottom heat: in short, they may be grown without having an Orchid house at all; any common stove will do, or even a well-heated pit. I have sold several collections this last season to gentlemen who have no other Orchids, but who have bought these for the express purpose of growing for decoration of the dinner table, on which they have a beautiful appearance under gas light. As table decoration, therefore, is in fashion, plants such as these will doubtless be much sought after for that purpose, and subjecting them to such treatment for a few hours in a warm room will do them little harm, provided they have glasses over them with a ventilator at the top to let out moisture so as to keep the leaves dry; being grown, as they generally are, in small pots, they can be plunged in handsome vases with ornamental glass tops. When done with, take them back into the heat. For room-work let the foliage be dry, and there should not be much moisture at the roots. I should not, however, advise them to be used in the way just described on a frosty night; but during mild weather it will not do them any more harm than taking them to a flower-show. When with C. B. Warner, Esq., at Hoddesdon, I showed *Anœctochili* at Regent's Park and Chiswick, under bell-glasses for years, and I never found them injured by such changes. I, however, always took care to prepare them before starting, by not giving them too much moisture, and putting them in a cooler house, which is the best way for all plants of a tender kind before taking them to a show. Few plants would be injured if treated in a proper way before leaving a warm house. The plants in question were shown for several years, and though that was more than six years ago Mr. Warner has some of the same stock in his collection at Stratford in fine condition: and these are grown without bottom heat and under bell-glasses in the shade. I mention this to show how many years *Anœctochili* may be grown when subjected to the treatment best suited for them. I think the stock at Hoddesdon has lasted sixteen years or more, and I have no doubt, if treated in the same way, it may keep on for years to come.

When out of health I have found that the best way is to turn them out of their pots and examine their roots; if in a bad state below, wash the bottom of the plant, and repot in fresh soil. Thus treated I have known them to improve and do well for some time, provided they have not been allowed to get too much out of order before being seen to; if so, there may be little hopes of all care bestowed on them being of any use. These plants, unlike some Orchids, have no thick woody pseudo-bulbs to support them; their small fleshy stems require constant attention to keep them in a thriving state, but with care they may be grown in perfection. The flowers are small and unattractive, and often injure the plants; I, therefore, always pinch them off when they appear, and doing so induces them to break freely.

The treatment they require is a warm house or pit, where

the temperature ranges in winter, by night, from 55° to 60°, or a few degrees higher would not do them any harm; the warmth may rise to 65° by day, and by sun heat to 70°; during March, April, and May the night temperature may range from 60° to 70°, and afterwards a few degrees higher will not be injurious. From March to October is the best time for growth, during which they will require a good supply of moisture at their roots; in fact, they should never be allowed to get dry,—if so, they will most likely perish; but from October to March only give them sufficient to keep the soil damp. They succeed best under bell-glasses or in a case in small pots, with a little air always on to keep them in a healthy condition. Some will do without glasses, if in a warm house. I grow them in sphagnum, chopped into small pieces, intermixed with a little good fibrous peat and silver sand, all well mixed together. I have found river sand to answer the same purpose. I have grown them in both with good drainage. These plants do not require large pots, as they do not make much root, they therefore succeed well in small pots; and if bell-glasses are used, plunge the pot into a large one so that the bell-glass fits the outside one. If grown in cases put them in small pots, and arrange them in the case by placing good drainage at the bottom and sphagnum on the top of the drainage, with some sand on the top of the moss to set the pots on; then arrange the kinds in the case so that the different colours may make a good effect. The pots should be perfectly clean, with good drainage at the bottom, covering with a little moss, and filling up with the material recommended. In placing the roots in the pot, raise the stem a little above the rim; after that pot once a year, and I have found, at the end of February or the beginning of March, the best time. The glass or case in which they grow should be kept perfectly clean, in order that they may have plenty of light, but not any sun, which is injurious to them.

They are propagated by cutting the plants into pieces just

below the first joint, and so as to have a root attached to each piece. For this purpose strong plants should be selected; and, in cutting, take care that the bottom piece has two eyes, one to root from, and the other to push into a shoot; place them in small pots in the material already named. The "bottom," or plant cut, should be put under a bell-glass, or placed in the case, where it will soon throw up a young shoot, which is best left on till well rooted; then cut it off from the old plant, and treat it like the portion first removed, leaving the old part in the pot, which will throw up again and form another plant from the bottom eye.

Anœctochili are subject to different kinds of insects; red spider is very destructive to them, and, if allowed to get ahead, soon spoils the foliage, but if constantly looked after it may be kept in check. Take the plants out of the case, and examine the under sides of the leaves, and rub them over with a sponge. I should not, however, do this except there is necessity for it. Thrips is another enemy which must be kept under in the same way, or by fumigation, taking the glasses off for a short time. Cockroaches, too, should not be allowed to get near them. I have seen great mischief done by them; they eat the young stems, and must be sought after by candle-light, or lay some Chase's beetle poison in different parts of the house. The latter is sold in boxes, and should be laid down about twice a week till the cockroaches are destroyed. In smoking, be careful not to give too strong a dose. The best way is to give it three times every other night till both thrips and spiders are destroyed.

Anœctochilus argenteus.—A handsome distinct free-growing species, with leaves two and a half inches long, one and a half broad, and having stems four inches in height; ground colour light green, with well-defined silvery markings. This does not require so much care as some others. I have seen it grown in a warm house without a bell-glass, and with a good supply of water at the roots in a shady place.

A. Bulleni.—This charming new species, from Borneo, grows six inches high, and has leaves two and a half inches in length; ground colour bronzy green, marked through the entire length with three broad distinct lines of coppery red, varying at times to golden stripes. This will prove, I believe, one of the best that has yet been introduced, as it seems to be very free in growth.

A. El Dorado.—A distinct species, and very difficult to cultivate. It appears to be a deciduous kind, and is often lost by people throwing it away, thinking that it is dead, whereas, if left, it will push up again; the foliage is dark green, with small tracery of a lighter colour. This plant should not be allowed to get too dry at the roots when at rest; if so, it will die; it requires a good deal of care to keep it in good condition.

A. intermedius.—A fine distinct species, and one of freer growth than some others. It grows three inches high, and has leaves two and a half inches long, one and a half inch broad, with a soft silky surface; colour dark olive, striped and veined with gold. This will do well without a glass, if in a warm house, and shaded from the sun.

A. Javanicus.—A species not so good as many others, but still worth cultivating on account of distinctness of colour; height four inches; leaves one and a half inch long, and one inch broad; ground colour dark olive-green, with blotches of a lighter green.

A. Lobbii.—A fine distinct species, attaining a height of three inches, and with leaves two and a half inches long, and one and a half inch broad; colour dark olive, with light marking over the whole surface.

A. Lowii.—A splendid species, the largest of the genus; it grows six inches high, and has leaves from four to five inches long, and three inches broad, beautiful and velvety; colour rich dark green, shading off to mellow orange brown, lined from stalk to point with well-defined deep golden veins, and crossed by lines of the same attractive hue. This remarkable plant was found by Mr. Hugh Low near an opening of a large cavern in the interior of Borneo.

A. Lowii virescens.—A charming variety of the above, growing equally large; foliage lighter green, with brighter markings over the whole surface.

A. maculatus.—A beautiful and distinct sort, growing five inches high, and having leaves three inches long, and one and a half inch broad; each leaf is edged with dark green, and has a handsome silver-frosted band down the centre. A very free-growing plant, and one which may be grown without a glass.

A. Nevilleanus.—A distinct and pretty species from Borneo; grows about three inches high, and has leaves an inch and a half long; ground colour dark velvet, enriched with blotches of orange. Apparently a free grower.

A. "petola".—One of the finest of the genus, very free-growing, and easy to increase. Of this there are two varieties; one not so good as the other, but both handsome. It grows four inches high, and has leaves from two to three inches long, and two inches broad, resembling light-coloured velvet, enriched with well-defined lines, and bands of a deep golden colour covering the whole surface. A magnificent species, which has only been in cultivation about three years.

A. querceticolus.—A distinct species, but one only worth growing for the sake of variety. Grows from three to four inches high, and has leaves two inches long, of a light green, with blotches of white down the centre. A free grower.

A. Roxburghii.—A pretty and distinct kind, which grows three inches high, and has leaves two and a half inches long, and one and a half inch broad, with dark velvety appearance; over the whole surface are well-defined lines of silver. Several kinds are grown for this species, but none so good as the true one, which is very rare.

A. Ruckerii.—A new and pretty species, just imported from Borneo. I have, therefore, not seen it sufficiently to say much about its merits. The plants are very small. Mr. Low, the importer of it, describes it as having leaves broadly ovate; ground colour bronzy green, with six rows of distinct spots running down the entire length of the leaves. It looks distinct from most others.

A. setaceus.—One of the handsomest of the genus, and one of the oldest. It grows four inches high, and has leaves two inches long and one and a half-inch broad; surface a beautiful velvet, veined in regular lines, and covered with a network of gold. There are several varieties of this charming plant, all of which are free growers.

A. setaceus cordatus.—A remarkably handsome variety; grows three inches high, and has leaves two inches long, and one and a half inch broad; resembles the former, but is rounder in the leaf, and the gold markings are broader. A rare variety.

A. setaceus grandifolius.—A beautiful kind, growing two and a half inches high, and having leaves two inches long, and one and a half inch broad; foliage light green, beautifully laced and banded with a network of gold. A rare variety, and one which I have only seen in the collection of J. A. Turner, Esq., Manchester.

A. striatus.—A distinct species; grows five inches high, and has leaves three inches long, of a dark green, with broad band of white down the centre. A free-growing plant, and one which may be cultivated without a glass in a warm house.

A. Veitchii.—A rare and fine species, named after its importer. It grows four inches high, and has leaves three inches long and two inches broad; ground colour beautiful light velvety green, interspersed from stalk to point by well-defined lines, and bars of the same colour, but lighter. A free-growing plant, and one which grows nearly as large as *Lowii*.

A. xanthophyllus.—A splendid species, and very distinct from any other variety in cultivation. It attains a height of five inches, and has leaves two and a half inches long, and one and a half inch broad; ground dark velvety, with broad orange and green stripes down the centre, and covered with a beautiful golden network. A free-growing species.

ANGRÆCUMS.

Curious Orchids, of which there are several species, but only a few are worth growing : these, however, are handsome in growth, and ought to be in every collection. In habit they resemble *Aerides*, having beautiful evergreen foliage, which, in some kinds, is regularly curved, and very graceful. The flowers are produced on long spikes from the axils of the leaves. Even when not in bloom, the plants themselves are objects of interest, and give a noble appearance to the house in which they are grown. The flowers are not so much prized as they ought to be, the spikes being stiff, and therefore comparatively unsuitable for exhibition; but blooming, as they do, in winter, they are invaluable to those who look for beauty at that season. They continue six weeks or more in perfection. If strong they generally flower every year, and also produce young plants at the bottom. The latter, if required, should be taken off when rooted; if not, leave them on, in order to make a finer specimen. They require the same treatment and material as *Aerides*, and, like them, are best grown in the East India House.

A. bilobum.—An elegant little Orchid from Cape Coast, with dark green leaves and very compact growth; the blossoms are white, and have a small tail about two inches long. The plant blooms from October to December, and remains two or three weeks in good condition. I have grown this plant on a block, but it does best in a basket, where the roots obtain plenty of moisture.

A. caudatum.—A singular free-flowering species, from Sierra Leone, with pale green drooping foliage, ten inches long, and very compact growth, producing racemes of flowers a foot or more in length; the flowers are greenish yellow, mixed with brown, the labellum being pure white, and furnished with a tail of pale green colour, about nine inches long. I have sometimes seen twelve or more of these curious flowers on a spike. Its season of blooming is from June to September, and it continues in perfection a long time. This has always been a rare plant.

A. eburneum.—A noble free-flowering Orchid from Madagascar, strong growing, with light green stiff foliage, eighteen inches long, very thick and broad; the flowers, which are of ivory whiteness, are produced on upright spikes eighteen inches long, and, if kept from damp, last four or five weeks in perfection. It blooms during the winter months, and is on that account valuable.

A. eburneum superbum.—A fine variety of *eburneum*, brought from Madagascar by W. Ellis, Esq., of Hoddesdon. It is stronger in growth than the former; the blooms are large, ivory white, coming about the same time as those of *eburneum*, and continuing a long time in beauty.

A. eburneum virens.—A free-flowering variety, the flower spikes of which are not so stiff and therefore are more graceful than *eburneum*. The blossoms are greenish white, and the plant has dark green foliage, about ten inches long. In perfection during December and January.

A. sesquipedale.—A wonderful plant, brought by W. Ellis, Esq., of Hoddesdon, from Madagascar, where he found it growing on trees. Foliage dark green, about ten inches long; blooms beautiful ivory white, and very large, with two green tails hanging from the bottom of the flower, about ten inches in length. In blossom in November, December, and January, and lasts three weeks in beauty. A very rare species, and certainly the finest of its class.

ANGULOA.

There are only five plants of this genus that I have seen. The flowers are large and beautiful. The plants make good subjects for exhibition, especially *A. Clowesii*, whose colour is different from that of many of our Orchids, making it valuable. The pseudo-bulbs are large, about three inches high, with broad flag-shaped leaves a foot or more long; they all produce their flowers, which are about six inches high, from the base of the bulbs just as they begin to grow. All the kinds are best grown in pots, with rough fibrous peat, good drainage, and plenty of heat and moisture in the growing season. The East India house is the most suitable place for them during their season of growth; afterwards they may be moved to a cooler place. They ought to have a good season of rest, and during this time they should be kept rather dry, till they begin to show signs of moving, when they must be grown in pots in peat and treated as already recommended. They are propagated by dividing the bulbs just before they begin to grow.

A. Clowesii.—A charming and free-growing species from Columbia; sepals and petals bright yellow; lip pure white. Blooms in June and July; lasts long in perfection if kept in a cool house.

A. Clowesii macrantha.—A fine variety, also from Columbia, growing about the same height as the preceding; the flowers, which are bright yellow, spotted with red, are produced in July, and, if kept dry, continue three or four weeks in perfection. A scarce plant.

A. Ruckerii.—A handsome Orchid from the same country as the former, and flowering at the same time. The sepals and petals rich brownish orange, lip greenish yellow:

lasts two or three weeks in good condition. Also a rare plant.

A. uniflora.—A good species from Columbia, flowers freely; the blossoms are white, and are produced in June and July, lasting two or three weeks in flower.

A. virginalis.—A pretty species, likewise from Columbia, which grows about a foot high, with dark green bulbs; the blossoms, which are white, spotted all over with dark brown, are produced in June and July, and last three weeks in bloom. A rare species.

ANSELLIA.

Two of this genus at least are well worth growing; both are noble free-flowering Orchids, growing about four feet high, and blooming in winter, when they produce large spikes of flower, which, if kept in a cool house, last long in perfection. *Ansellias* require good-sized pots, as they root very freely, and are of easy culture, provided they get the heat of the East India house while growing, and a good supply of water at the roots; be careful, however, not to wet the young growths, as that might cause them to rot. *A. Africana* was found in Fernando Po, at the foot of a palm-tree, by the late Mr. John Ansell. All of them are propagated by dividing their bulbs after they have finished their growth, or just after they have done blooming.

A. Africana.—A free-flowering noble Orchid, producing upright stems from three to four feet high, with light evergreen foliage; the flowers are pale yellow, spotted all over with dark brown; lip yellow; keeps in beauty for two months. I have seen upwards of one hundred flowers on one spike.

A. Africana gigantea.—A very fine variety of the preceding, producing upright spikes from the top of the bulbs; it flowers about the same time and is of the same colour as *Africana*, lasting a long time in perfection; very rare.

ARPOPHYLLUM.

There are only three plants belonging to this genus with which I am acquainted that are worth growing. All three are of handsome habit, with graceful evergreen foliage, and having beautiful upright spikes of flowers, charmingly arranged, looking not unlike rows of small shells clustering round the spike a foot or more in length. *A. giganteum* makes a fine exhibition plant, whose colour is distinct from that of most Orchids. All of them require the heat of a Cattleya house, and they are best grown in pots, in peat and good drainage, with a liberal supply of water at the roots when growing. They are propagated by division.

A. cardinale.—A very beautiful species from Guatemala, with dark evergreen foliage. Flowers produced on upright spikes a foot high; sepals and petals light rose; lip deep red; in bloom during the summer months, and lasts three or four weeks in perfection.

A. giganteum.—A magnificent species from Guatemala, and certainly the best of the genus, having dark evergreen foliage and a graceful habit. The flower-spikes, which are produced from the top of the bulbs, grow from about eight inches to a foot high; the blossoms are beautiful dark purple and rose: they are produced during April and May, and last three weeks in perfection. Previously to being exhibited, this requires to be kept at the coolest end of the house, for it will generally come in too early if not kept back.

A. spicatum.—A pretty evergreen Orchid from Guatemala, with dark red flowers on an upright spike, which continues in beauty three or four weeks during the winter months.

BARKERIAS.

These plants are deciduous, losing their leaves during their season of rest; they are small-growing, but free in producing flowers, which are both rich and delicate in colour. These plants merit a place in every collection. I have seen only five species. The *Barkerias* are compact-growing, with upright slender bulbs, from the top of which the numerous flower-stems are produced. These plants are best grown on blocks of wood of a flat shape, so that the plants can be tied on the top without any moss. They send out their thick fleshy roots very freely, and will soon cling to the blocks. They require to be grown in a cool house, where they can receive air every day during their season of growth. The Mexican house will be the most suitable place for them, but during their season of growth they require a good supply of water. Twice a day in summer will not be too much for them; but during their season of rest very little water will suffice,—only enough to keep their bulbs from shrivelling, about two or three times a week. They should be suspended from the roof, near the glass, where they can receive plenty of light, but not too much sun.

B. elegans.—A splendid species from Guatemala, producing upright spikes during the winter season; sepals and petals dark rose; lip reddish crimson, spotted and edged with a lighter colour; flowers as large as those of *B. spectabilis*. The finest of the genus, and very rare; of this there are two varieties, one not so good as the other.

B. melanocaulon.—A pretty and free-flowering Orchid from Costa Rica. The sepals and petals are lilac pink, the labellum with a spot of green in the centre. It produces its blossoms on an upright spike from June to September, and will continue in perfection a long time. A very rare and desirable species.

B. Lindleyana.—From the same country. The flowers are produced on a long spike, and are of a rich purple colour, with a blotch of white in the centre of the lip. Blooms in September and October, and lasts long in good condition. A scarce plant.

B. Skinneri.—A beautiful free-flowering Orchid from Guatemala, with deep rose-coloured blossoms, which are produced on a spike sometimes two feet long, with as many as twenty to thirty flowers on each spike. It will continue in flower from November to February. A valuable Orchid for winter-blooming.

B. spectabilis.—A charming species from the same country as *Skinneri*, the flowers of which proceed from the top of the bulb, on a spike bearing eight or ten flowers of a rosy pink, or blush, dotted with deep crimson. It blooms in June and July, and lasts three or four weeks in perfection, if kept in a cool house. This makes a splendid plant for exhibition. I have seen specimens at the Chiswick and Regent's Park shows, with as many as twenty spikes on one block. A very distinct and desirable Orchid.

BLETIAS.

These are terrestrial Orchids. There are several species of this genus, but there are only a few that are worth growing. They are of easy culture, and may be grown in any warm house, so long as frost is kept from them; the bulbs are round and flat, from which proceed long narrow leaves. They are deciduous. The best material for growing them in is loam and leaf-mould mixed together, with about two inches of drainage in the bottom of the pot, covered with a layer of moss or rough peat: then fill the pot with the mould to within an inch of the top, place the bulbs on the top of the mould, and cover them over. They require a good supply of water in the growing season, but not much heat. After their growth is finished, give them a good season of rest; and they should be kept rather dry till they begin to grow.

B. campanulata, from Peru. The blossoms are of a deep purple, with a white centre: flowers at different times of the year, and lasts long in perfection.

B. Shepherdii, from Jamaica. The flowers are purple, marked down the centre with yellow : blooms during the winter months on a long spike, which keeps in perfection three or four weeks.

B. patula, from the same place as the last. Produces its dark purple flowers, which last three or four weeks, on a long spike, in March or April. These plants require to be well grown to make them flower. The colour is distinct from many of our Orchids, and the flowers produce a good effect in a house. These plants are not thought much of by many Orchid-growers, but I think they are worth the care that is bestowed upon them on account of their colour.

BOLBOPHYLLUM.

There are several species of this genus, but only one that is worth growing, *Henshalli;* the flowers of the others are curious, particularly the labellum, on which the least breath of air or the slightest motion causes a tremulous or dancing movement. They are chiefly valued as curiosities, require but very little room, and thrive best on small blocks of wood with a little moss, suspended in a warm part of the house; the roots require a good supply of water. They are propagated by dividing the bulb.

B. barbigerum, from Sierra Leone. A curious dwarf species, sepals and petals greenish brown; the lip is covered with dark-coloured hair. It lasts long in bloom.

B. Henshalli, from Java, introduced by Messrs. Veitch of Exeter and Messrs. Rollisson of Tooting. The flowers are large, the sepals deep yellow, the upper part marked with purple and spotted. It produces its solitary flowers during the summer months, and lasts long in beauty.

B. saltatorium.—A curious dwarf Orchid from Africa, of a greenish brown colour. Blooms at different times of the year, and lasts some time in perfection.

BRASSAVOLA.

There are several species of this genus, but only a few that are showy and worth growing. They are of easy culture, and grow best in a little moss on blocks of wood, suspended from the roof. They require a liberal quantity of water during the growing season, but afterwards they need watering less frequently. They are best grown in the warmest house, and are propagated by dividing the plants.

B. acaulis.—A very good species from Central America, with rush-like foliage and compact growth. Flowers large, creamy white, produced in September, and remain a long time in beauty.

B. Digbyana.—A fine compact evergreen species from the West Indies. The plant is about six inches high, the sepals and petals of the flower are creamy white; lip the same, streaked with purple down the centre. It produces its solitary flowers during the winter months from the top of the bulb. Lasts about two or three weeks in bloom.

B. glauca.—A desirable compact evergreen Orchid from Mexico. Its blossoms are creamy white, with a pink mark on the upper part of the lip. It blooms in February or March, and lasts two or three weeks in perfection. This is rather difficult to flower in some collections, but it does bloom every year if the plant is strong, producing one large flower from a sheath at the top of the bulb.

B. venosa.—A pretty free-flowering species from Central America, small and compact, with white flowers, which are produced at different times of the year. It grows best on a block suspended from the roof.

BRASSIAS.

These are not thought much of by many Orchid-growers, but there are a few kinds that may be recommended. They are rather large-growing plants, of easy culture, and will do either in the East India or a cooler house. The flowers are produced from the side of the bulbs on a long drooping spike. They are all evergreen, with good foliage, each leaf being a foot or more in length. They are best grown in pots, with rough fibrous peat and good drainage, and require a liberal supply of water at the roots in the growing season; afterwards just enough water to keep their bulbs plump will suffice, for they should never be allowed to shrivel. They are propagated by dividing the plants when they begin to grow.

B. Lanceana.—A free-flowering Orchid from Demerara, blooming at different times of the year, and bearing yellow blossoms spotted with brown, which last three weeks in perfection. There are two varieties of this plant, one much better than the other, having larger and brighter coloured flowers.

B. Lawrenceana, from Demerara, blooming abundantly from June to August. The colour of the flower is yellow and green spotted with brown, and lasts three or four weeks in good condition, if kept dry.

B. maculata major, from Jamaica. A free-flowering Orchid, sepals and petals greenish yellow spotted with brown; lip white spotted with dark brown. Flowers in May and June; lasts five weeks in bloom if kept in a cool house.

B. verrucosa.—A curious species from Mexico. The upper part of the flowers is of a pale green; the lip white, marked with green warts. It blossoms abundantly in May and June.

B. verrucosa superba.—A fine variety from Mexico, whose growth is stronger than, and the flowers twice the size of, the preceding, and of a lighter colour. This is the best of the genus I have seen, and is well worth a place in every collection.

B. Wrayæ.—A very good species from Guatemala, producing its flowers on spikes two or three feet long; sepals and petals yellowish green, blotched with brown; the lip is broad and yellow, spotted with brown. Blooms from May to August, and continues flowering for two months.

BROUGHTONIA.

B. sanguinea, from Jamaica; the only one of the genus that I have seen cultivated. A very compact evergreen-growing plant, which succeeds best on a block of wood, with a little moss, suspended from the roof. It requires a good supply of heat and moisture in the growing season, and produces its spikes of crimson flowers from the top of the bulb during the summer months, lasting a long time in good condition. This plant ought to be in every collection, on account of the distinct colour of its flowers. It is propagated by dividing the plant.

BURLINGTONIA.

There are some beautiful species of this genus; they are very compact in growth, except *amœna*, with beautiful evergreen foliage, from four to six inches high. They produce their delicately-coloured flowers on drooping spikes shooting from the sides of their bulbs. These plants ought to be in every collection, however small, as they require but very little room, and may be easily grown to perfection. I find them thrive best in baskets with sphagnum moss and potsherds, and a good supply of heat and moisture while growing. They require but little rest, and should never be suffered to get too dry at the roots. They are propagated by dividing the plant. The following are among the best kinds of this beautiful class of plants with which I am acquainted.

B. amœna.—A beautiful free-flowering species from Brazil, but a straggling grower; it keeps on growing and flowering and throwing out roots all up the stem. The flowers, which are produced on upright spikes, are of a delicate white pencilled with light rose. They are produced during the winter months. It succeeds best in a basket or on a block suspended from the roof.

B. candida.—A handsome free-flowering Orchid from Demerara, producing drooping spikes of flowers, which are of a delicate white, with the exception of the upper part of the lip, which is yellow. It flowers at different times of the year. This plant has been very scarce till within the last two years. Thomas Bewley, Esq., of Black Rock, Dublin, in whose collection I saw some large masses growing on blocks, has, however, imported a fine lot. Mr. O'Brien, his gardener, informed me that to grow them in anything like perfection they must have a good supply of moisture at the roots, and should never be allowed to get dry.

B. fragrans.—A charming Orchid from Brazil. The white flowers, with yellow down the centre of the lip, are produced on a spike in April and May, and last three or four weeks in beauty, if kept free from damp.

B. Knowlesii.—A beautiful new species, somewhat similar in habit to *venusta*, very dwarf and compact. The flowers are white, in long racemes, slightly tinged with a pinky lilac. It blooms during the autumn, and continues in perfection a long time. A scarce Orchid.

B. venusta.—A very good species from Brazil, the blossoms being white, with yellow down the centre of the lip. It blooms at different times of the year; lasts two or three weeks in good condition.

CALANTHE.

There are some beautiful species belonging to this class of plants, which are great favourites, and ought to be in every collection. They are of easy culture, having handsome evergreen foliage, except *vestita* and *Veitchii*, which are deciduous, losing their leaves during the season of rest. All their flowers are striking, and generally attractive. Most of them are rather large, upright-growing plants, some of the leaves being a foot and a half long and six inches broad. Their long spikes of flowers rise from the bulbs, and come up between the leaves. They generally make their growth after the flowers have faded. These are terrestrial Orchids, and are best grown in pots of a large size, with loam, leaf-mould, and rotten dung, mixed together. When they are planted, two inches of drainage should be put at the bottom of the pot, then a layer of moss on rough peat; after which the pot should be filled up with the mould, and the plant left about level with the rim. These plants are best grown in the Indian house, and require well watering at the roots in their growing season, so that the mould is never allowed to get dry. They require but little rest; and during that time not so much water, only enough to keep the soil slightly damp. These plants are very much subject to the brown and white scale, which should be diligently sought; for if allowed to accumulate, the plants will not thrive. They are propagated by dividing the plant. The following are the most beautiful of this class. There are several others; but as many of them are not worth growing, I notice only those that are good.

C. Dominii.—A good and distinct hybrid, which grows in the same way as *masuca*. It was raised by Mr. Dominy from seed in this country, and is a cross between *masuca* and *veratrifolia*. A fine free-blooming plant.

C. furcata.—A showy Orchid from India, which is very free in producing its spikes of cream-white flowers, three feet long, and lasting in perfection six weeks. It flowers in June, July, and August, and is a good plant for exhibitions.

C. masuca.—A magnificent and free-flowering Orchid from India, producing its flowers on a spike two feet long. Sepals and petals deep violet colour, with a rich purple lip. It blooms in June, July, and August, and lasts six weeks in perfection. This is a charming plant for exhibition, the colour being very distinct.

C. masuca grandiflora.—A charming variety, with evergreen foliage and flower-spikes from three to four feet high; continues blooming for three months; sepals and petals white shading off to lilac, lip bright lilac. One plant of it was shown last year at five shows, proving it to be a fine variety for exhibition, on account of its long continuous blooming season. Very scarce.

C. Veitchii.—A beautiful mule, raised between *Limatodes rosea* and *Calanthe vestita* by Mr. Dominy. It is a deciduous plant, and grows like *C. vestita;* the flower-spikes are eighteen inches high; the blossoms a rich rose colour. This will prove a useful species.

C. veratrifolia.—A noble species from India; its spikes of flowers, which are of a delicate white, frequently attain the height of two or three feet; it blooms freely from May to July, and will continue blooming for two months. The flowers should be kept free from damp; if they get wet they are apt to become spotted. This also makes one of the finest exhibition plants we have; it is a very old plant, but no collection ought to be without it.

C. vestita rubra oculata.—A charming free-flowering Orchid from Moulmein; deciduous, and producing, from October to February, long drooping flower-spikes, which have a white, downy covering, and rise from the base of finely-formed and silvery-green bulbs when the latter are destitute of leaves. The sepals and petals of the flowers are of a delicate white; the lip is the same, with a blotch of rich crimson in the centre. No collection ought to be without both varieties. They are fine plants for winter blooming, the flowers being two inches across. We have a plant of this fine species with thirty spikes, and frequently with from twenty to thirty flowers on each spike, continuing in perfection for three months.

C. vestita lutea.—A variety of the preceding with white sepals and petals, and a lip of the same colour, with a blotch of yellow in the centre; its flower-spikes are produced in the same way as the last, and at the same time; it is nearly equal to it in point of beauty, and though a deciduous plant is useful for winter decoration.

CAMAROTIS.

Camarotis purpurea.—A beautiful upright-growing Orchid, with leaves three inches long throughout the whole length of the stem. It produces its flower-spikes, which are about eight inches long, from the side of the stem; the blossoms are rose coloured, and appear from March to May, lasting two or three weeks in beauty. This plant requires care to grow it well; it may be planted either in a pot or basket with moss, and requires a good supply of heat and moisture over the roots and leaves during the period of growth, but needs very little rest, and should never be allowed to shrivel. The East India house is the most suitable place for it. A fine specimen of this species was shown at the Chiswick and Regent's Park exhibitions in 1850. This single plant, on which there were more than 100 spikes of flowers, was grown by Mr. Basset, gardener to R. S. Holford, Esq., Tetbury, Gloucestershire.

CATTLEYAS.

These rank among our finest Orchids; they are general favourites, and there can be little doubt that, as the mode of treatment which they require, and the ease with which they may be brought to a high state of perfection become better understood, they will be extensively cultivated. Many of the bulbs are singular and agreeable in form; and the dark evergreen foliage of the plants when in a healthy condition, together with their compact habit of growth, renders them peculiarly attractive. Some of the sorts have only a single leaf at the top of each bulb; others, as *C. Skinneri* and *C. intermedia*, have two, and *C. granulosa* and *C. Leopoldiana* have three : the flowers are large, elegant in form, and scarcely surpassed in their brilliant richness and depth of colours ; the most frequent of which are violet, rose, crimson, white, and purple, with their intermediate shades. The flower-scape, which is enclosed in a sheath, rises from the top of the bulb, and a single spike sometimes contains as many as nine perfect flowers ; and I have seen as many as thirty. As soon as the flowering is over, the *Cattleyas* generally begin to make their growth for the next season ; but some of them, as *C. Walkeriana, C. violacea, C. superba,* and several others, flower while making their growth. In the cultivation of *Cattleyas* I have found them thrive best in pots, with the exception of *C. Walkeriana, C. marginata,* and *C. citrina,* which grow best on blocks with a small quantity of moss. If room be not abundant, all the kinds may be grown on blocks, but they will require more attention, and seldom thrive so well as in pots. They require a good depth of drainage. I generally fill the pot about half full of pot-sherds, which I cover with a layer of moss, and then fill up the remaining space with peat, taking care always to have the plants elevated above the rim of the pot.

I grow all the *Cattleyas* in the Mexican house, and am accustomed to give them a good supply of heat, and not too much water, while they are growing. Water applied to the roots once or twice a-week will be sufficient for those in most vigorous state of growth : too much water is apt to cause the bulbs to rot. So long as the soil remains moist no water is required ; and when the soil becomes dry, water should be applied to the roots, not to the bulbs, as it is apt to injure them greatly. When the plants have made their growth they should be allowed to rest, and be kept rather dry, giving them just water sufficient to prevent their shrivelling.

A long season of rest is very advantageous to the plants, causing them to flower more freely and grow more vigorously afterwards. Those plants growing on blocks will require a good supply of water at their roots every day during summer, and twice or three times a week in winter.

Cattleyas require to be kept perfectly clean and free from insects. They are subject to the white scale, which should never be allowed to accumulate, as it is then difficult to remove, and the plants are in danger of being destroyed. In order to keep the plants free they should be carefully looked over every three or four weeks, and sponged with clean water of the same temperature as that of the house. They are propagated by dividing the plants.

The following list comprises the most beautiful and valuable species of this splendid genus; and the collection in which they are found will seldom, if ever, be without one or more of the sorts in flower.

Cattleya Aclandiæ.—A charming Orchid from Brazil, of a dwarf habit, bulbs seldom being above six inches high. It is a very shy-blooming plant; but its large chocolate-coloured flowers are variegated with yellow, and have a rich rose-coloured lip. It flowers in June and July, and remains long in perfection. There is a fine specimen of this, about two feet high, in the collection of J. Day, Esq., Tottenham, on a block of wood.

C. amabilis.—A magnificent free-flowering Orchid from Brazil, growing about eighteen inches high, making two growths in one year, and blooming from the one that is formed in spring. On each spike are from three to five blossoms, which remain about four weeks in perfection; the sepals and petals are of delicate pink; the lip is large and of the richest crimson. Undoubtedly the finest of the intermedia class, which it closely resembles; it blooms during the summer months, and makes a fine exhibition plant. Very rare.

C. amethystiglossa.—A beautiful and distinct new species from Brazil, and one of the finest I have seen; grows from two to three feet high in the way of *Leopoldii*, with two leaves on the top of the bulb; from the centre of the leaves are produced spikes, with six or seven flowers on each, measuring more than five inches across; sepals and petals light rose-spotted with rich purple; lip deep purple; blooms in March, April, and May, and will last five weeks in perfection. I have only seen this in the collection of R. Warner, Esq., of Broomfield; I know of no one else who has it true: plants sometimes sold for it appear to differ from it both in growth and flower.

C. bicolor.—A beautiful and free-flowering Orchid from Brazil; sepals and petals pale green, with a rich purple lip. It blooms in September, and remains a long time in perfection. It produces as many as eight or ten flowers on a spike. This is a strong grower, rising eighteen inches or two feet high. There are two varieties of this plant, one much better than the other.

C. candida.—A desirable free-growing species from Brazil, about a foot high; sepals and petals are of a delicate white, slightly shaded with pink: the lip is of the same colour, with a shade of yellow in the centre. It flowers from July to November, and lasts three or four weeks in good condition, if the flowers are kept free from damp. This plant makes two growths in a year, and blooms from both, producing three or four flowers on a spike.

C. citrina.—A beautiful dwarf plant from Mexico, with bright yellow flowers, one or two together, and large for the size of the plant; blooms from May to August, and lasts two weeks in perfection. This is best grown on a block of wood, and the plant should be tied to the block with the leaves hanging downwards, as it is found growing beneath the branches of trees in its native country.

C. crispa.—A splendid free-growing Orchid from Brazil, about a foot and a half high; it flowers in July and August: the blossoms are pure white, with a rich crimson stain in the middle of the lip. A single spike frequently produces four or five flowers, which continue in perfection for two weeks. This is a fine plant for exhibition in July. The colour of the flowers is remarkably attractive, and always produces a good effect in a collection.

C. crispa superba.—A magnificent variety, the flowers of which are larger than those of the preceding; sepals and petals pure white; lip rich crimson and beautifully fringed. I consider this to be one of the finest *Cattleyas* in cultivation if true, but many are sold under this name which have no claim to it. It blooms in July and August, and will last from two to three weeks in beauty.

C. Edithiana.—A most splendid species from Brazil, with dark green foliage a foot high; in growth like *C. Mossiæ;* the flowers are very large, measuring in diameter from six to seven inches; sepals and petals light mauve; lip also mauve striped with white, upper part buff. The flowers are produced in May and June, and remain in perfection three or four weeks; the only example I have seen of this, which will make a fine exhibition plant, is in the collection of R. Warner, Esq., at Broomfield.

C. elegans.—A most beautiful species; sepals and petals pale purple, suffused with cinnamon brown; labellum (form of *C. Loddigesii*) of the most brilliant purple; flowers in clusters after the manner of *granulosa*. This beautiful species was imported by Messrs. James Backhouse and Son, of York.

C. granulosa, from Brazil.—A free-growing species, producing large olive-coloured flowers, with rich, brown spots; the lip is whitish, spotted with crimson. It blooms in August and September, and remains long in perfection. This species is not so good as many of the *Cattleyas;* but where there is room it is worth growing, for its distinct colours.

C. guttata, from Brazil, is a free-growing Orchid, about twenty inches high. The flowers are of a greenish yellow, beautifully spotted with crimson; the lip is white, stained with purple. This plant, when well grown, will produce as many as nine or ten flowers on a spike. It blooms in October and November, last two weeks in bloom and is a very distinct species.

C. guttata Leopoldii.—A charming variety from Brazil; grows about twenty inches high, and has short dark green foliage; a free-blooming kind, producing its flowers after it has made its growth. Sepals and petals dark brown, spotted with crimson; lip purple. I have seen this with thirty flowers on one spike, a condition in which it was exhibited at Regent's Park by Mr. Page, gardener to W. Leaf, Esq., Streatham, in whose collection it was grown; it was the finest spike I ever remember seeing; it generally has from six to ten flowers on a spike, and continues in perfection about three weeks; makes a fine plant for exhibition on account of its distinct colour. Of this there are six fine varieties in the collection of T. Robinson, Esq., Birkenhead, all very fine and worth describing; but unfortunately I did not take any notes of them when in bloom.

C. Harrisoniæ, from Brazil.—A free-growing Orchid, attaining the height of twenty inches; blooms in abundance from July to October; the flowers will last in good condition three weeks: the blossoms are of a beautiful rose colour, having on the lip a slight tinge of yellow. This is a noble plant, and amply repays the care required to grow it well. A fine specimen of *C. Harrisoniæ,* grown by Mr. Woolley, late gardener to H. Bellenden Ker, Esq., Cheshunt, and having more than fifty flowers opened at once, was shown at the Regent's Park Exhibition in 1851.

C. Harrisoniæ violacea.—A charming Brazilian variety, which grows about two feet high, and makes two growths in one year, flowering on both in July, August, and September, and will last in bloom four weeks if kept in the coolest house. The sepals and petals are of a beautiful violet; lip same colour, with a little yellow in the centre. Will make a fine exhibition plant, the colour being very different from that of many other Orchids.

C. hybrida.—This was raised in this country by Mr. Veitch, a pretty species, but not so good as many others; still well worth growing.

C. intermedia violacea.—A beautiful Orchid from Brazil; a free-grower, and about a foot high. Sepals and petals are delicate rose colour, and the lip has a rich purple spot in the centre. It blooms in May and June, and lasts three or four weeks in good perfection, if kept in a cool place. This species frequently produces as many as nine flowers on a spike, and makes a fine plant for the May and June shows. There are several varieties of *C. intermedia.*

C. intermedia superba.—A splendid variety from Brazil; grows about fourteen inches high, and makes two growths in one year; but only flowers from the one made in spring. Sepals and petals delicate rose; lip broad, and of the richest purple; from four to six blossoms produced on a spike; makes a fine plant for exhibition, and will last about four weeks in perfection.

C. labiata.—One of the best of the *Cattleyas;* comes from Brazil; is a free-blooming species, and grows about twenty inches high. The flowers are rose coloured, with a rich crimson lip. It blooms in October and November, and will continue in perfection three or four weeks. The flowers are very large, often six inches across, with three or four on a spike. The finest specimen I ever saw was in the collection of R. Hanbury, Esq., Pole's Gardens, near Ware. The plant was cultivated in a large tub, about two feet across. This superb *Cattleya* was first flowered by the late Mr. Cattley, of Barnet, in honour of whom the genus is named.

C. labiata atropurpurea.—A splendid variety which grows about the same height as the preceding; flowers large, and of a much richer colour ; sometimes produced in fives on a spike. This I have seen in the collection of E. McMorland, Esq., Haverstock Hill.

C. labiata pallida.—A beautiful variety from Brazil, whose growth is shorter than that of the two preceding ; leaves upright, and of a lighter green ; blooms in August. Sepals and petals light pink ; lip crimson, beautifully fringed ; a useful variety on account of its flowering earlier than *labiata*, and, if the flowers are kept dry, it will remain three weeks in good condition.

C. labiata picta.—When well bloomed one of the finest of all *Cattleyas;* flowers large, but produced sparingly, and often deformed ; when well expanded, however, magnificent ; grows about a foot high, with light green foliage ; a single bloom sometimes measures seven inches across. Sepals and petals pure white ; lip of the richest crimson, and beautifully fringed. Flowers during June and July, and will continue three or four weeks in perfection. This was first flowered at Sion House, and is often called Sion House *Cattleya*.

C. Lemoniana.—Distinct from *Mossiæ* in growth, the leaves being shorter. It grows about a foot high, and has light green foliage ; the flowers are produced during the summer ; colour pale pink, with yellow in the centre of the lip ; will last three or four weeks in bloom.

C. lobata.—A charming species from Brazil, very much like *crispa* in growth, but shorter in both bulb and leaf. Blossoms deep rich rose ; a very shy-flowering plant, producing its blooms in May and June, and continuing about three weeks in perfection. One seldom sees this plant exhibited on account of its shy flowering : it was, however, shown last year with several flowers on it, and had a good effect, its rich colour being attractive.

C. Loddigesii, from Brazil.—Grows a foot or more high ; a free-flowering species ; sepals and petals of a pale rose colour, tinged with lilac; the lip is a light rose, marked with yellow. It blooms in August and September, and remains long in perfection, producing three or four flowers on a spike, and is a desirable species.

C. marginata.—A handsome dwarf plant from Brazil, about six inches high. Sepals and petals rosy crimson; lip deep rose, margined with white. It produces its bloom in September and October, and lasts three weeks in perfection. The flowers are large, one and sometimes two being on a spike. It is a very desirable plant, and ought to be in every collection, however small, as it requires but very little room. It grows the best on a block of wood, with a little sphagnum moss, suspended from the roof of the house, with a good supply of water at the root in the growing season. A scarce Orchid.

C. maxima.—A magnificent species from Columbia; grows from a foot to eighteen inches high. A distinct growing plant, producing its blossoms, which are bright rose, in November and December, four or five on a spike; lip richly variegated, with dark crimson veins down the centre. Of this there are two varieties. J. Day, Esq., of Tottenham, and J. A. Turner, Esq., of Manchester, both possess a very fine variety of this charming plant, which will last in perfection three weeks.

C. McMorlandii.—A fine species from Brazil; in the way of *Mossiæ*, growing about a foot high, with dark green foliage. Flowers about six inches in diameter. Sepals and petals beautiful light rose; lip yellow and fringed. Blooms in June and July, and remains three or four weeks in perfection. Very distinct, and makes a fine exhibition plant.

C. Mossiæ.—This magnificent and free-flowering Orchid is from La Guayra. It grows a foot or more high, blooms from March to August. The flowers are very large, a single one being four or five inches across, and three or four growing on one spike. There are many varieties of this plant, with blossoms of every shade of colour, from pale pink, white, crimson, and rose purple. This species makes a noble plant for exhibition. I have seen specimens of *C. Mossiæ* at the Chiswick and Regent's Park shows, with thirty or forty flowers upon a single plant. The flowers will remain in perfection for three or four weeks, if kept in a cool place.

C. Mossiæ aurantiaca.—A splendid variety from Brazil; grows about a foot high. Sepals and petals rose pink; lip darker in colour, with rich orange yellow in the centre. Of this there is a fine variety in the collection of J. A. Turner, Esq., Manchester, called *Aurea grandiflora;* it continues in flower same time as *Mossiæ*.

C. Mossiæ superba.—A magnificent variety from La Guayra. I have seen this with flowers measuring seven inches across. The leaves are broad, and the plant grows about a foot high, sepals and petals, splendid rose; lip large, of rich crimson, and beautifully fringed. A fine plant for exhibition on account of its free-flowering habit and rich colour.

C. pumila.—A charming species, from Brazil, about six inches high. It flowers in September: the blossoms are rose coloured, with a crisped lip of a crimson colour, often edged with white. It remains three or four weeks in perfection, and is like *marginata* in growth.

C. quadricolor.—A very rare and pretty species, which grows about ten inches high, and produces its blossoms in May and June on the young growths. Sepals and petals light rose; lip same colour, yellow on the upper part.

C. Schilleriana.—A charming species, which grows much like *C. Aclandiæ;* the foliage is however darker and rounder; blooms during the summer months from the young growths; flowers large, and nearly the same colour as *Aclandiæ*, but much darker, and they remain in perfection three or four weeks, if kept dry. Of this there is a very beautiful variety called *Regnelli*.

C. Skinneri.—A beautiful and free-flowering plant from Guatemala; grows about a foot high, and blooms in March,

April, and May: the blossoms are rose-purple, and remain three weeks in perfection, if the flowers are kept dry. This fine species of *Cattleya*, when grown strong, will produce as many as nine or ten flowers on a spike. It is one of the finest Orchids that can be grown for any of the exhibitions in May, the colour being distinct and different from any of the other *Cattleyas*.

C. superba.—This truly beautiful plant comes from Guyana. It is a slow-growing species, and usually ten inches high. It flowers in June and July. The blossoms, which are of a deep rose colour, with a rich crimson lip, will remain in bloom three weeks: it produces three or four flowers on a spike, and is a very rare and distinct species.

C. violacea.—A handsome, free-flowering species from Brazil, which grows about twenty inches high. It produces flowers in abundance, from July to September: the flowers last in perfection three weeks: the blossoms are of a beautiful, deep violet-rose colour, and frequently present four or five flowers on a spike. This plant makes two growths in one year, and blooms from both. It is a fine exhibition plant for the late shows in July.

C. Wagnerii.—A splendid free-flowering *Cattleya* in the way of *C. Mossiæ*, and with flowers equal in size; sepals and petals white; lip also white, with rich yellow in the centre. A fine plant for exhibition, producing blossoms in June and July, and lasting about three weeks in perfection.

C. Walkeriana.—A truly elegant dwarf species from Brazil, about four inches high, with large, light, rose-coloured flowers; the lip, which is a richer rose than the other part, having a slight tinge of yellow: generally two flowers on a spike, five inches across. It blooms twice a-year, February and June, on the young growth: the blossoms last four or five weeks in beauty, which is longer than that of any other *Cattleya*: it is also sweet-scented, and will fill the house with perfume. I always grow this plant on a block of wood, surrounded by a little sphagnum moss, and suspended from the roof in a place where there is plenty of light, but not too much sun; very rare species.

C. Warnerii.—One of the finest of all *Cattleyas;* grows in the same way as *labiata*, and with blossoms equal in size, a very useful species for summer exhibition; flowers large, more than six inches across; sepals and petals, beautiful rose; lip large, of a rich crimson, and finely fringed. This rare plant I saw for the first time in the collection of R. Warner, Esq., and I thought it the finest and most magnificent *Cattleya* that had ever come under my notice; of this there are, however, several varieties.

C. Warscewiczii.—A magnificent species which grows about a foot high, with light green foliage, in the way of *Mossiæ*. The flowers are large; sepals and petals purplish white; lip rich crimson. Of this there are also several varieties; blossoms during the winter months, and continues three or four weeks in perfection. Mr. Baker, gardener to A. Bassett, Esq., Stamford Hill, had a splendid variety of this, with many blossoms on it. At present it is a great rarity.

C. Warscewiczii delicata.—A magnificent variety in the way of *Mossiæ*, but still very distinct: grows about a foot high: blossoms six inches across; sepals and petals white; lip large, with a beautiful yellow centre, and a tinge of rose, and white on the outside. Blooms in December and January: very useful during winter, and continues in perfection three or four weeks. This fine variety was exhibited before the Floral Committee of the Royal Horticultural Society in February, 1862, by E. McMorland, Esq., Haverstock Hill.

CHYSIS.

This is a beautiful, though limited, class of plants, there being only four of this genus with which I am acquainted. They are deciduous, losing the leaves during their season of rest. The bulbs are thick and fleshy, and about a foot long, producing their flowers with the young growth: these will do on blocks of wood, but grow much finer in baskets or pots, with peat, moss, and potsherds. They require a liberal supply of heat and moisture in their growing season, but after they have finished their growth should be moved into a cooler house till they begin to grow: then they may be taken back to the India house. During their season of rest they require but very little water, and are propagated by dividing the plants just as they begin to grow.

Chysis aurea.—A charming Orchid from Venezuela, producing its flowers on a short spike, generally twice a-year: the blossoms are yellow, the lip is marked with crimson. It flowers at different times of the year, and lasts two weeks in beauty.

C. bractescens.—A fine species from Guatemala: its flowers are produced on a short spike, sometimes six together, each flower measuring two or three inches across; the sepals and petals are white; the lip has a blotch of yellow in the centre. It blooms in April and May, lasts two or three weeks in perfection: it makes a good show-plant.

C. lævis, from Guatemala.—The blossoms are cream-coloured, with a blotch of yellow on the lip: it flowers in the same way as the two former kinds, but at different times of the year, and lasts two weeks in good condition. A scarce plant.

C. Limminghii.—A charming species from Guatemala, which grows a foot high, producing delicate pink and rose-coloured flowers very freely along with the young growth in May and June, and they continue in perfection three weeks. This makes a good exhibition plant, but requires a cool house to keep it back for late flowering.

CŒLOGYNE.

There are many species of this class of plants, some of which are very beautiful; the colour of the flowers of some being rich and of a most delicate hue, whilst others are not worth growing. The following produce the best flowers, and all I have seen are evergreen. Almost all the *Cœlogynes* have bulbs from three to six inches, from which the flowers proceed with the young growth: they make their leaves after they have done blooming: the flowers of some are very large, measuring as much as three inches across.

These plants are all best grown in pots, with peat and moss. Some of them will do on blocks, but the pot-and-peat culture is the best. They require good drainage, and an abundance of water at their roots in their growing season, and are best grown in the East India house; but after they have finished their growth a cooler house will do. During their period of rest they should have but little water. They are propagated by dividing the bulbs.

Cœlogyne cristata.—A magnificent Orchid from Nepaul: a dwarf evergreen species, leaves six inches long; the flowers proceed from the bottom of the bulb, on a drooping spike, six or eight together, each flower being three or four inches across: the colour is a beautiful white, with a blotch of yellow on the lip. It blossoms in February and March, and will continue in perfection four or five weeks if the flowers are kept free from damp. This is the finest of the genus, and ought to be in every collection. I have seen it with as many as sixty blossoms on it at one time. This occurred on a specimen grown by Mr. Baker, gardener to A. Bassett, Esq., Stamford Hill.

C. Cummingii.—A pretty species from Singapore. Sepals and petals white; lip bright yellow, with white down the centre. It lasts long in beauty.

C. Gardneriana.—A very good Orchid from the Khoosea Hills; grows about a foot high; the flowers, which are white tinged with yellow, are produced on a drooping spike. It blooms during the winter months, and lasts three or four weeks in good condition.

C. Lowii.—A fine evergreen species from Borneo : a rather large-growing plant, about two feet high : the flowers are produced on a drooping spike, about a foot long, twelve or more flowers together, which are large, of a pale yellow and chocolate colour. It blooms in June or July; lasts two weeks in good condition. This requires a large pot to grow it in perfection, being one of the freest of all the *Cœlogynes.*

C. media.—A pretty small growing species, with short round bulbs, and leaves seven inches long ; flowers on spikes, ten inches high ; sepals and petals, creamy white ; lip, yellow and brown ; blooms during winter, and lasts in perfection three or four weeks; grows well on a block, or in a pot in peat.

C. pandurata.—A curious and distinct flowering Orchid from Borneo ; grows eighteen inches high, and has flat shiny bulbs, from the bottom of which the flower spikes proceed, several together ; upper part of the flower green, lip nearly black.

C. plantaginia.—A desirable Indian species, with greenish yellow flowers, having a white lip streaked with brown ; a distinct kind.

C. speciosa.—A free-flowering plant from Borneo, evergreen, and about eight inches high ; the flowers are creamy white with a dark brown lip, and are produced on a short spike, two or more together, at different times of the year. It lasts long in bloom.

CORYANTHES.

These are very large, extraordinary-looking flowers. Before the blossoms open they are in the shape of a Chinese foot ; after opening, they form a sort of a cup, having above it a pair of fleshy horns, from which a sort of liquid exudes and drops into the cup. They produce their flowers from the bottom of the bulbs on a spike, four or five together. The plants are evergreen, with leaves ten inches long, on short bulbs three inches high. The leaves are two or three inches broad. They will grow either in baskets or pots, with moss and peat and good drainage; they require a liberal supply of water at the roots during their period of growth, with a good heat. After they have made their growth they should be kept rather dry, so that their bulbs be not allowed to shrivel. They are propagated by division of the bulbs.

Coryanthes macrantha.—The best of the genus comes from the Caraccas, and flowers in May, June, and July: lasts but three or four days in bloom; the colour orange-yellow, spotted with purple: the lip is red. This plant ought to be in every collection.

C. maculata.—From Demerara.—Flowers dull yellow, spotted with dull crimson; blooms during the summer months, and lasts but three days in beauty.

C. speciosa.—This singular Orchid also comes from Demerara, is about a foot high, with pale yellow-coloured flowers, which are produced in April or May: lasts three or four days in bloom.

CYCNOCHES.

Another singular tribe of plants, whose flowers are of a very peculiar form, being in the shape of a swan. They are not thought much of by many growers of Orchids, but some are well worth growing; they are of easy culture, and produce their flowers freely; all are deciduous, and lose their leaves as soon as they have finished their growth. The bulbs, which are thick and fleshy, are from six to ten inches high, and have three or four leaves on the top of each bulb. They produce their flowers, which are large, from nearly the top of the bulb, some of them several together. They are best grown in the East India house, in pots, with rough fibrous peat and good drainage, with a liberal supply of water at the roots in their growing season; afterwards they may be kept much cooler, and should be placed near the glass, to receive all the light possible ; and during this time they must be kept rather dry, having only just enough water to keep their bulbs from shrivelling. When they begin to grow, move them back into heat. They are propagated by dividing the bulbs when they begin to grow.

Cycnoches barbatum, from New Granada.—A dwarf species, sepals and petals greenish white, spotted with pink ; lip the same colour. It blooms in June or July : lasts in flower two or three weeks. A very scarce plant.

C. chlorochilum.—A very good species from Demerara ; the flowers, which are of a yellowish colour, are produced in June or July, and last three weeks in good condition if kept dry.

C. Loddigesii.—A very curious Orchid from Surinam. It produces its blossoms on a spike, six or eight together, which are four inches across ; the sepals and petals are of a brownish-green colour, with darker spots, and bearing resemblance to the expanded wings of a swan. It blooms in July and August, and continues in perfection three weeks. A desirable species, which ought to be in every collection.

C. pentadactylon.—This is also a very curious species from Brazil ; the flowers are very large, the colour being yellow and brown. It flowers at different times of the year, and lasts long in beauty.

C. ventricosum.—A free-flowering Orchid from Guatemala ; the blossoms are greenish-yellow, with a white lip : blooms in June, July, or August, lasting in bloom three weeks.

CYMBIDIUM.

' There are several species of this genus, but only a few that are good ; some of them are very handsome, and delicate in colour. They are all evergreen, with beautiful foliage ; some are large-growing plants with short bulbs, from which the leaves and flowers proceed. They are generally free-flowering plants, and some of them produce pendulous spikes as much as two feet long. They are best grown in the East India house, in pots of large size; they require plenty of pot room, as they send out thick fleshy roots very freely. I grow these in rough peat with good drainage, and a liberal quantity of water at the roots during their period of growth; afterwards less water will suffice, and they should be kept at the cool end of the East India house. They are propagated by dividing the bulbs. The following are the best I have seen :—

Cymbidium eburneum.—A remarkably handsome species, the finest of the genus, with graceful foliage, and of very compact growth. The sepals and petals of the flowers are pure white: the lip is the same colour, with a blotch of yellow in the centre: the flowers are erect, about six inches high, and very large. It blooms in February and March, and lasts a long time in bloom. A very scarce plant. We are indebted to Messrs. Loddiges for the introducing of this magnificent plant, of which there are two varieties. I saw a pretty kind grown by Mr. Stone, gardener to J. Day, Esq., Tottenham. This is smaller in growth than *eburneum*, and the flowers are not so large as the former. Sepals and petals pure white; lip same colour, with a blotch of yellow in the centre, and rose-coloured spots on each side.

C. giganteum.—From Nepaul.—This is not so good as the others, but it makes a good plant for winter-blooming: a rather large-growing species; its brown and purple flowers are produced on long spikes during the winter, and last long in perfection if they are kept dry.

C. Mastersii.—A pretty Orchid from India. It is a great deal like *eburneum* in its growth, but the flowers are very different; they are produced on a spike, and are white with a yellow centre. This plant blooms during the winter, and continues long in flower.

C. pendulum.—A very good Orchid from Sylhet: a large-growing plant, with long drooping spikes from one to two feet long. Sepals and petals brown; the lip red, striped with white. It blooms in July or August, and lasts long in beauty.

CYPRIPEDIUMS.

These are all beautiful in foliage as well as in flower, and are worth a place in every collection. They are of easy culture, and require but little space; the form of the flowers is curious, being that of a slipper—they are generally called the Lady's Slipper, and are all dwarf, compact, and evergreen, the leaves of some being beautifully spotted. They produce their flowers from the centre of the leaves, on an upright stalk, and rise from six inches to a foot high. All are best grown in the East India house, except *insigne*, which thrives best in a cooler place, and will do well in a warm greenhouse. I grow all in pots with peat, loam, and sand; they all require a liberal quantity of water at their roots during their period of growth. They need but little rest, and should not be allowed to get too dry at the roots. The plants are not like many of our Orchids; they have no thick fleshy bulbs to supply them with nourishment. They are propagated by dividing the plant.

Cypripedium barbatum.—A pretty species, with beautifully spotted foliage; the colour of the flowers brownish-purple and white: it produces its solitary flowers at different times of the year, lasting six weeks in bloom. There are two varieties of this plant, one being much brighter in colour than the other.

C. barbatum grandiflorum.—A charming variety, the flowers of which are larger than those of any of the other kinds, and it has finely variegated foliage. Flowers in July and August, and continues six weeks in bloom.

C. barbatum superbum.—A fine variety, which grows in the same way as *barbatum*; but the foliage is more variegated, and the flowers much handsomer, the lip being very dark and the upper petal having more white. There is another variety called *nigrum*, with a larger lip, which flowers at the same time, and will last about six weeks in perfection. This makes a fine plant for exhibition.

C. biflora.—A handsome species from India, in the way of *barbatum*, but with more variegated foliage. Grows four inches high. The blossoms are produced on a spike ten inches long; sometimes two flowers appear on one stem; the top petal is very handsome, the upper part being a beautiful white, the other part of the flower a purplish brown. Blooms in February and March, and will keep six weeks in good condition. A rare plant.

C. caudatum.—A remarkable and curious Orchid from Peru, with light green foliage, producing its pale yellow and green flowers, several together, on a single spike; there are two tails, which hang downwards from each flower, about twenty inches long. It blooms in March, April, and May, and lasts three or four weeks in perfection.

C. caudatum roseum.—A variety which grows in the same way as the preceding, and flowers at the same time. Blossoms rose, intermixed with yellow and green. Well worth growing.

C. Dayii.—A charming plant, the foliage of which is beautifully variegated; very distinct from any other of the variegated class. Flowers in May and June, and lasts a long time in perfection. Blooms large. Sepals white, with green veins; petals purplish, tinged with green.

C. Farrieanum.—A beautiful and distinct species from India, with leaves three inches long, of a light green colour. Blooms very freely during the autumn months, and will continue in perfection six weeks. Petals white, striped with green and purple; lip large brownish green and purple.

C. hirsutissimum.—A beautiful Indian species, with pale

green foliage ten inches long; the flowers proceed from the centre of the young growths on a stem ten inches high; the blossoms, which often measure six inches across, are of a purple, light green, and brown colour: they open in March, April, and May, and last six weeks in perfection.

C. insigne.—A good old species from Sylhet, with light green foliage. Sepals and petals yellowish green—the upper petal is tipped with white, and spotted with brown; the labellum is orange and brown. It produces its solitary flowers during the winter, lasting six weeks in bloom.

C. insigne Maulei.—A charming variety of the preceding. Grows in the same way, and flowers at the same time. Upper petal, however, more white, and the whole flower better in colour. This was exhibited by Messrs. Maule and Sons, Bristol, after whom it is named. The blossoms will continue in perfection six weeks.

C. Lowii.—A curious and beautiful Orchid from Borneo, with light green foliage. This rare Orchid produces its flowers on a spike, two or more together, during the summer, and continues in perfection for two or three months if kept in a cool house.

C. purpuratum.—A pretty species from the Indian islands, with beautifully spotted foliage; produces its flowers during the winter months: the flowers are very much like *barbatum*, except the upper petal, which has more pure white on the end of the petal.

C. Schlimii.—A rare and beautiful species from South America. Very distinct in growth, as well as in flower, from most others; foliage eight inches long; light green; spikes branching, with as many as eight flowers on each. Flowers two inches across. Sepals and petals white and green; lip white, beautifully mottled, and striped with dark rose. This is a difficult plant to cultivate: it therefore requires great care. Pot in peat and sand, with good drainage, and take care that water does not lodge in the heart of the plant.

C. villosum.—A desirable Indian species, which grows about a foot high, of a light green colour, spotted on the lower part of the leaves with dark spot. Flowers on single stems, often measuring five inches across, and having a fine glossy appearance over their whole surface, which is orange red, intermixed with light green and dark purple. Blossoms during April and May, and continues six weeks in perfection. This makes a fine exhibition plant on account of its long continuance in flower and distinct colour.

DENDROBIUMS.

This is a magnificent class of plants; some of their flowers are very large and delicate in colour, and others delightfully fragrant. There are not many plants that surpass the *dendrobiums.* Their beautiful flowers are of nearly every shade of colour, and many produce them very freely; some are compact in their growth, while others are straggling; some are very gracefully growing plants, especially when cultivated in baskets and suspended from the roof, so that their pendulous bulbs hang down and exhibit their flowers to the best advantage : all the drooping kinds require to be grown in this way; some of them are evergreen, others are deciduous.

Though some of the *dendrobiums* require different treatment from others, they may all be successfully cultivated with proper attention. They are generally found in the hotter parts of India, growing on the branches of trees, frequently such as hang over streams of water; and to grow these plants to perfection they must have a good season both of rest and growth. They require to be grown strong to flower well. Some of the sorts are best grown in pots, with peat and good drainage; others do best on blocks of wood. I keep them all in the East India house. During their period of growth give them a good supply of heat and moisture, with a liberal quantity of water at the roots: in fact, the moss or peat never should be allowed to get dry while they are in a vigorous growth. After they have finished their growth, allow them a good season of rest by moving them into a cooler house, and during the time they are in a cool house give them but very little water, only enough to keep their bulbs from shrivelling. This is the only way to make them grow strong and flower freely; when they begin to grow they should be moved back into heat, and treated as is described above. They generally begin to grow after their flowers are faded.

They are propagated in different ways: some of them form plants on the old bulbs, which should be cut off and potted: some are propagated by cutting the old bulbs from the plants after they have done blooming; others are increased by dividing the plants according to the directions given in reference to propagation (p. 27). The *dendrobiums* are a large class of plants; some of them not worth growing, excepting for botanical purposes. The following are among the finest in cultivation, and will amply reward the care of the cultivator.

Dendrobium aduncum.—From Manilla.—A rather straggling grower: an evergreen species, producing its flower-spikes from the old bulbs. The blossoms, which are small, are white and pink in colour, and appear at different times in the year, lasting two or three weeks in perfection. This is best grown in a pot with peat and good drainage: it is not so good as many of the *dendrobiums*, but it is worth growing where there is plenty of room.

D. aggregatum majus.—A pretty, dwarf, evergreen species, from India: grows about four inches high, producing its pale yellow blossoms on a spike from the middle of the bulb: flowers in March and April, and lasts two weeks in bloom.

This will do either on a block, or in a pot with peat, and is a desirable species.

D. alba sanguineum.—A charming new Orchid from India: a compact-growing plant, with bulbs a foot high, and produces its large flowers, which are creamy white, from nearly the top of the bulb, with a crimson blotch on the centre of the lip. It blooms in July, and lasts a long time in perfection, and is best on a block, with moss. A scarce Orchid, and a fine showy plant for exhibition. This is a difficult plant to grow, so much so, that few people do it well. The best plant I have seen of it was in the collection of T. Robinson, Esq., Birkenhead, where it was growing on a block of wood, in sphagnum, and seemed to be quite at home, for the bulbs were very strong; Mr. Atkinson, the gardener, informed me that it required plenty of water during the growing season, and that it liked the warmest house.

D. album.—From India.—It produces its white blossoms during the winter. This is not so handsome as many, but it is worth growing where there is room.

D. anosmum.—A magnificent species from Manilla, which grows in the way of *macrophillum*, but has better shaped flowers, and no rhubarb scent; the flowers which are large, are produced in pairs down the stem; colour beautiful rose, and they last two or three weeks in perfection. Even now this is a scarce plant, it blooms during the spring months, and is best grown in a basket, as it is drooping and deciduous.

D. calceolarea.—A beautiful evergreen species from India. This is a large-growing plant, about four feet high when well grown: its flowers proceed from the top of the old bulb, on a raceme, twelve or more together, which are large; sepals and petals bright yellow, lip same colour; it blooms during the summer months, lasting but four days in bloom. This is best grown in a pot, with moss or peat.

D. Cambridgeanum.—A remarkably handsome Orchid from India: a deciduous species, of a drooping habit, about a foot long. The flowers are produced on the young growth in March and April: the sepals and petals are of a bright orange; the lip has a crimson blotch in the centre: it lasts in beauty two weeks, and is best grown in a basket, with moss, suspended from the roof.

D. chrysanthemum.—A charming species from India, deciduous and pendulous, growing about three or four feet long, and generally flowers along the stem at the same time it is making its growth. The colour of the flowers is a bright yellow, with a dark red spot on the lip: it blossoms at different times of the year, lasting two weeks in perfection, and requiring the same treatment as *Cambridgeanum*.

D. chrysotoxum.—From India.—An upright-growing plant, a foot or more high, and evergreen. It produces its pale yellow flowers on a spike from the top of the bulb; blooms during the winter months, and lasts two or three weeks in flower. Of this there are two varieties, one much better than the other: pot and peat culture suits it best.

D. clavatum.—A remarkably handsome species from India; grows two feet high, evergreen, and produces spikes of flower from the top of the bulbs; colour bright yellow, with a crimson spot in the centre of the lip; this is one of the finest of the yellow kinds, and it continues in perfection three or four weeks; makes a fine exhibition plant; pot culture in peat suits it best.

D. crepidatum.—A splendid deciduous drooping species from India; the bulbs grow a foot or more long; it is best grown in a basket, or on a block; sepals and petals white, tipped with pink; lip stained with yellow; blooms in April and May, and continues three weeks in perfection: a very rare species, and one which will make a good plant for exhibition if grown to a good size.

D. cretaceum.—A compact, deciduous-growing plant: it produces its white flowers during June and July, lasting six weeks in perfection. This will do either in a pot or basket, with peat or moss.

D. Dalhousianum.—This is a beautiful Indian evergreen species; bulbs are elegantly marked, and grow three or four feet high: it blooms from the old growth, with numerous flowers on a spike, in April and May. The flowers are large, of a pale lemon colour, with a pink margin, and two dark crimson spots in the centre: it lasts four or five days in beauty. This will grow either in a pot or basket, with moss. Specimen plants of this are scarce, and justly prized by those who possess them: makes a good plant for exhibition.

D. densiflorum.—A magnificent evergreen Orchid from India; compact and free flowering, of upright growth, a foot or more high. It produces its beautiful spikes of rich yellow flowers from the side of the bulb, near the top. It blooms in March, April, or May, and lasts from four to six days in perfection, if kept in a cool house. This is one of the showiest Orchids in cultivation, and is best grown in a pot with peat, and is one of the choicest plants we have for exhibition, on account of its colour. I have seen this with nearly one hundred flower-spikes on it at one time, in the collection of J. Day, Esq., Tottenham.

D. densiflorum album.—An Indian kind, distinct from the former, and called Schroder's variety; grows same height as *densiflorum*, and has foliage of the same colour; flowers pink and white, lip yellow; blossoms in April and May, and continues about ten days in perfection; it grows best in a pot in peat.

D. Devonianum.—This is one of the finest of the genus from India; a deciduous and pendulous-growing species. The blossoms proceed from nearly the whole length of the bulb, which sometimes attains the extent of four feet: the flowers are two inches across; sepals cream colour, shaded with pinkish purple; the petals are broader than the sepals, and are pink, with a deep purple stain; the lip is broad and fringed, rich purple on the ends, with two spots of rich orange on the column. It blooms in May or June, lasts two weeks in perfection, and is best grown in a basket with moss. This makes a splendid plant for exhibition.

D. Falconerii.—A magnificent Orchid from India, and one of the most beautiful of the genus; it is a pendulous grower, and so difficult to cultivate that only a few have succeeded well with it. Mr. Baynes, gardener to R. Michols, Esq., Bowden, Manchester, manages this plant better than any I have seen; he grows it on a block, without moss, merely tying the plant on with the bulbs hanging down; during the growing season it likes a good supply of water, but after it has finished growth, which is by autumn, it should be kept dry till it begins to show flower, merely giving just sufficient moisture to keep it from shrivelling; under this treatment it flowers every year, well repaying any trouble that is taken to induce it to produce its charming white and purple flowers, which open in May and June, and remain in perfection two weeks; a very rare plant.

D. Farmerii.—A beautiful, compact, evergreen Orchid, from India, with dark green foliage. This grows and flowers in the same way as *densiflorum*. The colour of the flowers is pink, with a yellow centre: it blooms in April and May, and lasts two weeks in beauty if kept in a cool house, requiring the same treatment as *densiflorum*.

D. fimbriatum.—A very good Orchid from India. A drooping, evergreen species; the bulbs grow two or three feet long, producing their flowers on a spike at nearly the end of the bulb: the flowers are of a bright yellow, and beautifully fringed. This plant will continue flowering from the old bulbs for years; it generally blossoms during the spring months, seldom lasting more than four days in bloom; is best grown in a basket, with moss.

D. fimbriatum oculatum.—A beautiful evergreen variety of the preceding, producing flower-spikes on the top of the bulbs, and growing from two to three feet high; it succeeds best in a pot, in peat; the flowers are large, of a rich orange yellow, with deep brown spot on the lip; blooms in March and April, if kept dry, and lasts ten days in perfection.

D. formosum.—A remarkably handsome, compact, evergreen Orchid; grows about a foot high, and blooms from the top of the bulb, three or four together: the blossoms are white, with a bright yellow centre, and frequently three inches across; they last six weeks or more in perfection. This may be grown either in a pot or basket with moss or peat: it makes a splendid plant for exhibition.

D. formosum giganteum.—A magnificent Indian variety of the preceding, much stronger in growth, and evergreen; the flowers, which are produced on the top of the bulbs, measure from four to five inches across; colour delicate white, with bright yellow on the centre of the lip; requires the same treatment as *formosum,* and remains in bloom about the same time. E. M'Morland, Esq., Haverstock Hill, possesses a fine specimen of this, with as many as twenty-one leading growths, which, when in bloom, must make a fine display.

D. Gibsonii.—A very pretty evergreen and upright growing Orchid from India, about two feet high: the blossoms are produced on the ends of the old bulbs; the sepals and petals are rich orange; the lip bright yellow, with two dark spots on the upper part: blooms during the autumn months, lasts two weeks in good condition. Same treatment as *formosum.*

D. Heyneanum.—A charming deciduous species from Bombay, which grows eight inches high, and produces spikes of white flowers from the tops of the bulbs at different times of the year, looking like sprigs of white May. It is best grown on a block, with a moderate supply of moisture, during the growing season. It continues about ten days in perfection.

D. Jenkinsii.—A beautiful dwarf evergreen species, from India, about two inches high : the flowers are very large for the size of the plant ; the colour is a pale buff, margined with yellow ; they are produced from the bulb, one or two together : lasts ten days in beauty. This is best grown on a bare block of wood, suspended from the roof.

D. lituiflorum.—A charming species from India, deciduous and pendulous : it succeeds best in sphagnum, in a basket suspended from the roof. It is difficult to grow, and requires great attention as regards moisture during the growing season, but it should be kept dry during winter. The flowers are large, and are produced in pairs up each side of the bulb. Sepals and petals dark purple ; lip white, edged with purple. Blooms in March and April, and will last two weeks in perfection.

D. longicornu majus.—A charming Indian variety, in the way of *formosum*, but with growth not so strong. Flowers white, except the lip, which has a yellow centre, and fringed. Of this there are two varieties, but the one now described is the best. It produces its blossoms from the top of the bulb in May and June, and they continue in perfection a long time if kept in a cool house. Treatment the same as that for *formosum*.

D. Lowii.—A fine new species from Borneo, where it was found by Mr. H. Low, to whom we are indebted for many beautiful Orchids. A very distinct plant ; it grows in the way of *D. longicornu*, with upright bulbs a foot high, furnished with dark green foliage, and producing flowers in dense racemes from the sides of the bulbs, seven together, two inches across. Colour bright yellow, with red markings on the upper part of the lip. This will succeed in a pot, or basket, or on a block, provided it has a liberal supply of water when in vigorous growth. Blooms in November.

D. macrophyllum.—A fine Orchid from the Philippine Islands, of pendulous habit; a deciduous species, losing its leaves just as it begins to show its flower-buds. The bulbs grow about two feet long, from which the flowers proceed in a row on each side; they are pink, tinged with rose colour, three or four inches across each flower, and lasting two weeks in perfection, if the flowers are kept dry. It is best grown in a basket, with moss. This makes a noble plant for public exhibition.

D. macrophyllum giganteum.—A showy variety from Manilla, which flowers in the same way as *macrophyllum*, and about the same time. Flowers from five to six inches across, delicate light rose, and they remain about two weeks in perfection. A fine plant for exhibition, but it requires to be kept in a cool house to keep it back for that purpose, as it generally blooms very early.

D. moniliforme.—A delicate species from Japan: an evergreen upright growing plant, about a foot or more high; blooms all up the bulb on two-year old growth, the colour being a light blue and white: it blooms during the winter months, lasting two weeks in beauty. This will grow in a basket, but I find it thrives best in a pot, with peat or moss.

D. moniliforme majus.—A fine Japanese variety of the preceding, growing the same height, but having larger flowers; the latter being four inches across, and richer in colour.

D. moschatum.—A handsome species from India: this grows in the same way as *D. calceolare*, the flowers are bright orange; lip chocolate, edged with yellow, and it lasts about as long in perfection.

D. nobile.—A magnificent Orchid from India; a free-flowering evergreen species, of upright growth: the blossoms, which are pink and white, with a spot of crimson in the centre of the lip, are formed along the sides of the bulbs. It blooms during the winter and spring months, lasting three or four weeks in good condition, if kept in a cool house. It will grow either in a pot or basket, with moss or peat. This is one of the finest exhibition plants we have.

D. nobile intermedium.—A pretty and distinct Indian variety, which grows in the same way as *nobile*, and flowers at the same time. Sepals and petals white; lip white, with a crimson spot in the centre. A desirable variety for winter decoration. A scarce plant.

D. nobile pendulum.—A fine evergreen variety of *nobile*, from India, and one which is best grown in a basket, on account of its pendulous habit. Flowers large, and richer in colour than those of *nobile*, and produced at the same time.

D. Paxtonii.—A remarkably handsome Orchid from India; an evergreen species, with upright growth, about three or four feet high : it blooms at different times of the year, from nearly the top of the old bulb, on a spike, and its flowers are of a beautiful orange colour, with a dark centre : lasts in perfection ten days, and requires the same treatment as *nobile*.

D. Pierardii.—A useful Orchid, for the winter and spring months, from India : a drooping, deciduous species, flowering in the same way as *macrophyllum*, and requiring similar treatment : its beautiful white flowers last three weeks in beauty.

D. Pierardii latifolium.—From India : the flowers are much finer than the old species, though of the same colour : it blooms in April and May. A scarce plant. I have seen it with seventy flowers on a single bulb. It requires the same treatment as the former, and is a useful plant for exhibition.

D. primulinum.—A beautiful free-flowering deciduous species, from India, of pendulous growth : the flowers, which are white and pink, are produced in two rows down the bulb in April and May, and they last in perfection ten days. It is best grown in a basket with sphagnum. A fine plant for exhibition. I have seen as many as sixty flowers on one bulb, when grown in baskets, and the bulbs drooping down. It has a very graceful appearance.

D. pulchellum purpureum.—A pretty, dwarf, deciduous species, from Sylhet; loses its leaves after it has finished growing, and generally begins to show flower in February all up the bulb: the sepals and petals are white, edged with green; the lip is beautifully fringed; it has a bright orange blotch in the centre: does well in a basket with moss; lasts two weeks in bloom.

D. sanguinolentum.—A good species from India, evergreen; the bulbs and leaves are violet or lilac coloured; grows three or four feet high; blooms from the end of the old bulbs, which keep blooming for years; sepals and petals fawn colour, tipped with spots of deep violet; the lip the same colour: it blooms during the summer and autumn months, and lasts two weeks in good condition: this will do either in a basket or pot, with peat or moss.

D. taurinum.—A vigorous growing magnificent species from India, often attaining a height of five feet, with upright bulbs. Flowers pure white, with the exception of the lip, which is margined with purplish violet, and the petals, which are reddish brown: it is best grown in a pot in peat. Of this a fine plant may be seen in the collection of S. Rucker, Esq., Wandsworth, under the care of Mr. Pilcher, the gardener.

D. transparens.—A beautiful small-flowering Orchid from India; blooms in the same way as *nobile;* the flowers, which grow in pairs along the bulb, are of a pale, transparent, pinkish lilac, stained in the middle of the lip with a blotch of deep crimson: it blooms in May and June, and does well grown in a pot with peat or moss. A rare plant.

D. tortile.—A charming species from Java; an evergreen, about two feet high; blooms in the same way as *nobile*, and requires the same treatment. The flowers are pale yellow, almost white: blooms in May and June, lasting a long time in perfection.

D. triadenium.—A delicate Orchid from India, of upright growth, and evergreen: produces its white and pink flowers on a small spike from the old bulbs, lasting two weeks in bloom: flowers at different times of the year, and is best grown in a pot with peat.

D. Wallichianum.—This is a beautiful Indian kind; has taller bulbs, much darker foliage, and richer-coloured flowers than *nobile*: it blooms at the same time, and requires similar treatment. This makes a noble plant for show.

DENDROCHILUMS.

A small genus of gracefully growing plants, of which I can speak only of two; both are small and compact in habit, and ought to be in every collection; they are evergreen, have small bulbs and narrow leaves, about six inches in length; their flower spikes, which are graceful and pendulous, are produced from the top of the bulbs, hanging down in interesting racemes; both sorts grow in the East India house in pots in peat, with good drainage; they like plenty of water during the growing season, but after they have finished growth less will suffice; they are propagated by dividing the bulbs just as they begin to grow.

Dendrochilum filiforme.—A charming species, which grows about six inches high. The flowers, which are produced in June and July, are of a yellowish green hue, and are prized for their gracefully drooping habit. When arrived at a good size, makes a nice exhibition plant.

D. glumaceum.—Another pretty species, with small evergreen foliage, and producing graceful spikes of greenish white flowers during the summer and winter months: its blossoms continuing three or four weeks in perfection.

EPIDENDRUMS.

Many of this large class of plants are scarcely worth cultivating, excepting for botanical purposes. Some, however, are sweet-scented. More growers of Orchids have been deceived in buying Epidendrums, than any other class of Orchids: the bulbs of so many kinds are so nearly alike, that it is difficult to tell what they really are till they flower. They may often be kept several years before they flower, and then, instead of something good, they produce frequently only dingy green flowers, about the same colour as the leaves. Some of these are very fragrant, and will perfume the whole house in which they are grown. There are, however, some beautiful species among this class, the flowers of which are very distinct in colour. The following comprise all the best sorts that I know. They are all evergreen, and compact in their habit, except *cinnabarinum, crassifolium* and *rhizophorum,* which are tall-growing, with long slender bulbs, and small leaves from top to bottom: the other sorts have short, round bulbs, with long narrow leaves, except *aurantiacum, bicornutum,* and *Stamfordianum.* These grow more in the way of the Cattleya, with upright bulbs, having two or three short leaves on the top. They all produce their flowers from the top of the bulb, except *Stamfordianum,* in which they rise from the bottom.

These plants will all do in the Mexican house, and may be grown on blocks of wood; but the pot-and-peat culture is the best. They require a season of rest, with the same treatment as the Cattleyas, excepting less heat: they are propagated by dividing the plants, as described in the remarks on propagation (p. 27).

Epidendrum alatum majus.—A pretty species from Mexico:

its spikes of flowers, produced in June or July, are pale yellow, with the lip striped with purple, and continue five or six weeks in beauty. There are several varieties of *alatum*, but I only know one, viz. the *majus* variety, which was flowered by Mr. Woolley, Cheshunt.

E. atropurpureum, var. roseum.—A fine Orchid, which grows twelve inches high, with evergreen foliage and short round bulbs, from the top of which it produces spikes of flower. Sepals and petals dark purple; lip large, of a beautiful rose colour, and continues in perfection a long time. It is best grown in a pot, in peat, with plenty of drainage.

E. aurantiacum.—A charming species from Guatemala. This plant grows in the same way as the *Cattleya Skinneri*, which the bulbs so nearly resemble as to be often mistaken for it. It grows a foot high, and produces its flowers from a sheath at the top of the bulb: the flowers are of a bright orange colour, with lip of the same, striped with crimson: it blooms in March, April, and May, lasting six weeks in perfection if kept in a cool house. There are two varieties of this plant, one of which never expands its flowers. I have them both growing in the same house, with the same treatment: the best variety opens its flowers, while the other keeps them nearly close;—a peculiarity which renders it not worth growing.

E. aloifolium.—A pretty species when well grown, from the same country as the former. This is pendulous-growing, and very distinct from any of the other Epidendrums, having curious and narrow-pointed leaves: the flowers, which are large, proceed from the top of the bulb, one or two together: the sepals and petals are greenish-yellow and brown, the lip is of a pure white; it blooms during the summer months, and lasts long in beauty. This plant is best grown in a basket with moss.

E. bicornutum.—A remarkably handsome Orchid from Guayana, about ten inches high: the flowers proceed from the top of the bulbs on a spike; sometimes on one spike there are as many as twelve beautiful flowers, each about two inches across, of a pure white, with a few crimson spots in the centre of the lip: blossoms in April and May, lasting two or three weeks in beauty. This is rather a difficult plant to grow: the best plant I ever saw was grown on a block of wood without any moss, and flowered five or six years in succession; but in the seventh year the plant seemed to lose its vigour, and never flowered afterwards, probably because the block began to decay and get sour. I have seen plants do well in pots with peat and good drainage.

E. cinnabarinum (from Pernambuco).—A tall-growing plant, four feet high, which blooms from the top of the bulb: the flowers are bright scarlet, and are produced in abundance in May, June, and July: it continues flowering for two or three months.

E. crassifolium.—This is not so good as many Epidendrums, but it is worth growing on account of its colour: it is a rather tall grower, two or three feet high; it produces its rose-coloured flowers in profusion in March, April, May, and June, and will continue blooming for three or four months, which makes it a valuable plant for exhibitions.

E. Hanburyanum (from Mexico).—This is not so showy as many of the Epidendrums, but is worth growing on account of its colour; sepals and petals deep purple, lip pale rose: blooms during the spring months, lasting long in beauty.

E. macrochilum.—A beautiful species from Guatemala; the sepals and petals brown, lip large, of a pure white, with a purple blotch at the base: it blossoms in April and May, and lasts five weeks in good condition, if the flowers are kept free from damp.

E. macrochilum roseum.—A beautiful variety of the former; the lip is of a darker rose colour, blooms at the same time, and lasts long in beauty.

E. maculatum grandiflorum.—A splendid Orchid, and certainly one of the best of the genus. Grows about twelve inches high, with short bulbs, and has evergreen foliage. Flowers produced on spikes fourteen together from the top of the bulb; sepals and petals creamy white spotted with black spots, lip pink: blooms in June and July, and will last several weeks in perfection. It is a very rare species, and makes a fine show plant.

E. phœniceum.—A fine species from Cuba; the sepals and petals purple; the lip of the same colour, mixed with pink and crimson: blooms during the summer months.

E. rhizophorum.—A pretty, but shy-flowering evergreen Orchid, and a straggling grower, being often ten feet high. It is best grown in a pot, in peat, with good drainage; when the plant becomes too tall, twist it round some sticks, which is the best way to make it flower; the latter, which are produced from the top of the bulb, are of a bright orange-scarlet, and the same spike will keep in beauty for three months. I have seen Mr. S. Woolley, of Cheshunt, have them in flower for twelve months.

E. Stamfordianum (from Guatemala).—A small-flowering species, producing its flowers in great abundance on a branching spike: they are of a greenish yellow, thickly spotted with brownish purple: it blooms in April and May, lasting a considerable time in perfection. There are two varieties of this plant; one has much brighter coloured flowers than the other; the best has longer and thinner bulbs than the other.

E. verrucosum.—A truly beautiful Orchid from Mexico; sepals and petals are of a delicate pink, spotted with crimson; the lip of the same colour: it produces its flowers in June and July, and continues flowering for four weeks. There are two varieties of this plant; one has much brighter coloured flowers than the other. A very scarce Orchid.

E. vitellinum.—A small-growing plant, but one of the finest of the genus, from Mexico : a very distinct Orchid; the sepals and petals rich orange scarlet, the lip a bright yellow : blossoms during the winter, lasting six weeks or more in good condition. A very scarce Orchid.

E. vitellinum majus.—The same colour as *vitellinum*, the only difference being in the larger size of the flowers, and in its blooming during the summer months; it lasts long in bloom. A very rare plant.

ERIOPSIS.

Eriopsis biloba.—A pretty species from South America, and the only one with which I am acquainted that is worth growing: it is evergreen and of upright habit, attaining a height of about ten inches; foliage dark green; flowers produced from the side of the bulb on a spike ten inches long; sepals and petals yellow and deep orange; lip white, spotted with dark brown, upper part orange. This succeeds best in pots in peat with good drainage, and it requires a liberal supply of water at the roots: it is propagated by dividing the bulbs. The coolest house will suit it.

GALEANDRA.

There are only three of this genus that I know of that are good and worth growing. They are deciduous and upright growing, with slender bulbs, producing their flower spikes from the top, just as they have finished their growth. These are best grown in pots, with peat, and good drainage, in the East India house, with a good supply of water at the roots during their period of growth. Afterwards they should be moved into the cooler house, placed near the glass, and kept rather dry, till they begin to grow, when they should be treated as before directed.

Galeandra Bauerii.—A desirable dwarf species from Guayana; the pink and purple coloured blossoms are produced on a drooping spike: blooms in June, July, and August; continues in perfection a long time. This makes a fine plant when well grown, and it is worth all the care that can be bestowed upon it.

G. cristata.—A good deciduous species from South America, which grows about eight inches high; the flowers, which are produced on a drooping spike, are of a pink and dark colour: they appear in July and August, and last four or five weeks in perfection.

G. Devoniana.—A beautiful slender Orchid from South America: it grows about two feet high; the blossoms, which are produced on spikes from the top of the bulbs, are white, beautifully pencilled with pink: it blooms at different times of the year, and remains a long time in beauty. A very scarce plant. I only know of two; one in the collection of S. Rucker, Esq., Wandsworth; the other in that of J. A. Turner, Esq., Manchester.

GOODYERAS.

An interesting class of plants, with compact habit, dark velvety-like foliage, marked with gold and silver lines down the centre of the leaves, and some of the kinds, like the genus *Anœctochilus*, have silver markings spread over the entire surface of the foliage. The plants have thick fleshy roots, and push from underground stems, forming beautiful round dwarf plants; the flower spikes which issue from the centre of the foliage attain a height of from six to ten inches; some of the kinds have delicate white flowers, especially *discolor*, which is one of the prettiest white winter flowering plants grown, very useful for bouquets, as well as for other purposes. Of this genus there are many fine species, of which we have only seen drawings, but I hope we shall soon be able to possess the plants themselves. The few I have seen, I will endeavour to describe. Some are of easy culture, while others are difficult to manage, but with care all may be had in perfection, which, when attained, amply repays any trouble that may have been bestowed. If grown in small pots and intermixed with *Anœctochilus*, they have a fine appearance, the dark foliage of the *Goodyeras* being very distinct from that of the other; they do not, however, require the same attention; on the contrary, they will do in any close house where there is a little warmth. I grow many of them in five and six-inch pots for blooming in the winter season, placing about six plants in a pot; they should be grown strongly, so as to ensure abundance of bloom; the soil I use is peat and sand, with a little loam, and I give a liberal supply of water at the roots during the growing season; they are propagated by cutting up the plants so as to have a piece of root attached to each piece; they may be grown where there is no Orchid House; a mixed stove suits them perfectly.

Goodyera discolor.—A beautiful plant from Brazil; grows about six inches high; foliage a beautiful dark velvety colour, with white marking through its entire length, flowers white, with a little yellow in the centre, and produced on a stem about ten inches high in winter, and they last a long time in perfection; a useful plant grown in five and six-inch pots, several plants being put in a pot so as to make a good show; even without any flowers the plants themselves are by no means unattractive.

G. Dominii.—A beautiful hybrid, raised from seed in this country, and as regards foliage, the best I have seen. I have not seen its bloom; the leaves larger than those of *discolor*, of a bronzy dark velvety-like appearance, with several prominent lines running their whole length, of a lightish colour—nearly white, and interspersed with smaller veins; a good addition to this class of plants contrasting as it does well with the *Anœctochilus.*

G. picta.—A distinct species, growing about three inches high; foliage one and a half inch long, light pea-green, with a paler band running through the entire leaf; I have not seen this flower, but it is worth growing on account of its foliage; a rare kind, whose roots are not so thick as those of *discolor;* it therefore requires more care.

G. pubescens.—A charming dwarf species; foliage green, enriched with white markings, in appearance something like *Anœctochilus argenteus;* this requires a cool house or pit to grow it in perfection; it is difficult to manage, and requires great care, being often destroyed by too much heat; grow it in pots not too large, giving a liberal supply of water during the growing season; in fact it should never be allowed to get dry at the roots. I have never seen this plant in bloom, but I grow it on account of its beautiful foliage.

G. rubro-venia.—A charming distinct species from Brazil in the way of *discolor;* grows several inches high, with velvet-like foliage, having three bands of red down each leaf; it has thick fleshy roots, and is of easy culture under bell glasses or in frames; I have grown it along with *Anœctochili* for several years on account of its foliage, but it will do in a pot with the same treatment as is usually given to *discolor*. A scarce kind.

GRAMMATOPHYLLUMS.

A small genus of Orchids, of which I only know two that are worth growing; of these one makes a magnificent specimen, having a noble palm-like appearance; it therefore requires considerable space to grow it in perfection. It is, however, a shy bloomer. I have seen plants grown for eight years without flowering; it has only once or twice been flowered well in this country—once by Mr. Scott, gardener to Sir George Staunton, and again by Mr. Carson, Nonsuch Park, Cheam; the latter had it very fine: this plant had been in the Cheam Collection for many years, but was sold about two years ago; it is now in the possession of S. Mendel, Esq., of Manchester, under the care of his gardener, Mr. Sharman, with whom I hope to see it flower again. The plants require to be well-grown, and after making a few strong growths, give a season of rest; *G. speciosum* requires to be grown in peat, in a pot of good size, and with good drainage; a liberal supply of water at the roots must also be given during the growing season, and it should be kept in the East India house; they are propagated by dividing the bulbs.

Grammatophyllum Ellisii.—A charming species brought from Madagascar by W. Ellis, Esq., of Hoddesdon, in compliment to whom it is named; it is a smaller growing species than *speciosum*, and more free-flowering, producing spikes of blossoms from the bottom of the bulbs along with the young growths; the flowers are large, of a yellow and brown colour, and they remain some time in beauty. This may be made to succeed on a block suspended from the roof, provided it has a plentiful supply of water.

G. speciosum.—A magnificent Orchid from Java, but, as I have just stated, shy-blooming; it grows from five to ten feet high, producing upright spikes from the bottom of the bulbs, which are very large; the flowers are also large, and of a beautiful yellow and brown colour : blooms during winter, and will last a long time in perfection if the flowers are kept dry.

HOULLETIAS.

A small class of Orchids, of which I only know two that are worth cultivating, and that chiefly on account of their distinctness of colour. They grow best in pots, in peat with good drainage, and like a liberal supply of water during the growing season; they are propagated by dividing the bulbs just before they begin to grow.

Houlletia Brocklehurstiana.—A distinct species from Brazil; grows eighteen inches high, and has short round bulbs, and flag-shaped, pale green leaves. The flower spikes are produced from the side of the bulbs; the blossoms measure from three to three and a half inches in diameter; petals, orange-brown, enriched with darker spots; lip yellow, and also spotted with dark brown.

H. odoratissima.—A good species from South America; grows eighteen inches high, and has light green leaves; the flower spike rises from the side of the bulb; blossoms two and a half inches across; petals, orange-brown, striped with a lighter colour; lip white, tipped with yellow.

HUNTLEYA.

This small genus of curiously-formed, though not very showy flowers, is of easy culture, and worth growing. The plants have evergreen foliage, about ten inches high, and are compact in their growth, with small bulbs, from which their flowers proceed. The blossoms are large, and stand about four inches high. These are best grown in pots, with peat and good drainage, in the East India house, with a liberal supply of water at the roots during the time they are in vigorous growth, but less afterwards. They require but a short season of rest.

Huntleya marginata.—A beautiful species from South America, which grows about ten inches high, and has evergreen foliage; the flowers are produced from the side of the bulbs; colour, pinkish purple and white; blooms at different times of the year, and continues a considerable time in beauty; a rare species.

H. meleagris.—The best of the genus, from South America. The flowers are yellow and brown; blooms in June and July, and lasts a long time in beauty. A scarce Orchid.

H. violacea.—A curious species from Guayana. It produces its solitary violet-coloured flowers at different times of the year; keeps in bloom four or five weeks.

II. Wailesiæ.—A singular dwarf plant, which produces white and purple flowers during the autumn months, lasting long in perfection; scarce, and very distinct.

IONOPSIS.

Ionopsis paniculatus.—A charming small, free-flowering Orchid, and one that ought to be in every collection; leaves, six inches high, with very small bulbs; the flower spikes which are branching, proceed from the axils of the leaf, and are about ten inches long; blossoms, pretty blush white, pencilled with light rose, and produced twice a year; succeeds best on a block, with a little live *sphagnum* moss round the roots, which require to be kept moist nearly all the year round. I have found it to do well in the coolest house suspended near the glass, where it continues in bloom for weeks at a time; a rare plant.

LÆLIAS.

This is a most lovely class of plants. Their flowers are large and very handsome, distinct in colour; most of them compact in their growth, with evergreen foliage, and resembling in many respects the genus Cattleya, to which some of them are equal in the beauty of their flowers. They produce their blossoms, on spikes of varied length, from the top of their bulbs. These plants merit a place in every collection, and will amply repay the cultivator for any care they may require. The Lælias are among our finest Orchids for winter and summer blooming. Some are best grown on blocks of wood with moss; others thrive well in pots with peat and good drainage.

The large growing kinds are best grown in pots, without

much water at the roots at any time, and require the same treatment as Cattleyas. Those on blocks require more water. All are best grown in the Mexican house, except *Perrinii*, which requires a little more heat to grow it in perfection. They are propagated in the same way as Cattleyas.

Lælia acuminata.—A pretty, delicate-flowering, and compact growing plant from Mexico; sepals and petals white; lip white, with a dark blotch on the upper part. It blooms in December and January, and lasts two or three weeks in beauty.

L. albida.—A lovely compact-growing species from Oaxaca; sepals and petals white; lip pink, with stripes of yellow down the centre. It blossoms in December and January, lasting a good time in beauty. There are two varieties of this plant. The best is called *superba*. In the latter, the flowers are much larger than in *albida*.

L. anceps.—A remarkably handsome Orchid from Mexico; sepals and petals rose-lilac; the lip of a beautiful deep purple. It blooms in December and January. The flowers are three or four inches across, and last a month in perfection, if kept in a cool dry house. Of this there are several varieties, one of which is called *Barkerii*.

L. autumnalis.—A lovely and showy Orchid from Mexico. It produces its blossoms on a spike twelve inches or more high, and often as many as nine flowers on a single spike. I have bloomed it with that number. The sepals and petals are of a beautiful purple colour; lip rose and white, with yellow in the centre. The flowers are four inches across: it blooms at the same time as *anceps*, lasting two weeks in good condition. There are several varieties of this plant, some of them much richer in colour than others; one fine one I saw grown by Mr. Toll, gardener to J. A. Turner, Esq., Manchester.

L. Brysiana.—A fine showy Orchid from Brazil, and very

distinct; grows like *Cattleya crispa*, with dark evergreen foliage, flowers large, produced three and four together during the summer months, sepals and petals beautiful light rose, spotted with a darker colour; lip dark crimson; lasting three weeks in perfection. *L. purpurata* is often grown for this species, but it is quite distinct in colour.

L. cinnabarina.—A charming distinct species from Brazil, very compact in growth; flowers reddish orange, produced on upright spikes, many together. It blooms in March, April, and May, lasting six weeks in beauty. This makes a good plant for exhibition on account of its colour.

L. elegans.—A magnificent species from Brazil, with evergreen foliage; grows about two feet high, and blooms at different times of the year; of this there are many varieties, and shades of colour varying from white to light rose, and pink, crimson, and carmine; in the true variety the sepals and petals are pale rose; lip brilliant purple: lasts about three weeks in perfection.

L. elegans, var. Dayii.—A splendid variety, which grows eighteen inches high, with two leaves on the top of the bulb of a light green colour; the plants flower twice a year; the blossoms are large; sepals and petals light rose, veined with a deeper colour; lip large, of a beautiful magenta colour—the top part light rose and throat yellow. I have only seen this in the collection of J. Day, Esq., Tottenham, in honour of whom it is named. It blooms in June and autumn, and lasts several weeks in perfection; it will make a fine exhibition plant.

L. elegans, var. Warnerii.—A magnificent variety of *elegans*, which grows about the same size; sepals and petals light rose; lip rich crimson: blossoms in June and July; will last in perfection three or four weeks; very rare, and one of the finest of its class, and is also a fine exhibition plant.

L. flava.—A very distinct species from Mexico; grows in

the same way as *cinnabarina*. The colour of the flowers is yellow: blooms at the same time as the former; lasts three weeks in perfection.

L. furfuracea.—A fine variety from Mexico; resembling *autumnalis* in growth, and attaining a height of ten inches, with light green foliage; flowers produced on upright spikes from the top of the bulb during autumn; colour dark purple; lip dark rose: blossoms individually five inches in diameter. This plant is somewhat difficult to cultivate, but I saw it doing well under the care of Mr. Baker, Gardener to A. Bassett, Esq., Stamford Hill, who has a very large example of it, and it blooms well with him every year; it is grown in a pot suspended from the roof of the Mexican house: a very rare plant.

L. grandis.—A splendid new species from Brazil, with pale green foliage; like *L. purpurata* it grows two feet high, and has several flowers on a spike; sepals and petals blush white, and it has a prettily marked palish lip which is more open and funnel-shaped than that of *purpurata*, the inside of the tube being distinctly visible; colour yellow, lined with purple, throat and mouth stained with rose and violet; in bloom during May and June, and lasts several weeks in beauty. A plant of this was exhibited by R. Warner, Esq., at the Royal Horticultural Society's exhibition last year, and received a first prize for a new plant; it is handsome and distinct.

L. Lindleyana.—Very distinct both in flower and growth; the foliage is like that of *Brassavola venosa*, and grows about eight inches long; sepals and petals rosy white, lip same colour, with the exception of the bottom part which is dark rose: blooms at different times of the year and continues in perfection six weeks.

L. majalis.—This is a glorious plant from Bolanos; a dwarf-growing species. The flowers are large, four inches across, of a delicate rose colour; the lip is striped and spotted with cholocate. It blooms with the young growth, and lasts five or six weeks in beauty. This is the finest of the genus, and is rather difficult to flower in some collections; but I have flowered it successively for several years. I keep it rather dry during the winter; a scarce plant.

L. Maryanii.—A pretty species, very compact in growth, with pale green foliage; sepals and petals flesh colour changing to salmon; lip mauve with buff stripes; blossoms during winter; a desirable species which continues in bloom four or five weeks at a time.

L. peduncularis.—A charming compact-habited evergreen plant from Mexico; grows like *L. acuminata;* sepals and petals beautiful dark rose; lip same colour, with darker spots in the centre: blooms during the winter months, and lasts two weeks in perfection; requires to be grown on a block.

L. Perrinii.—A truly beautiful species from Brazil; grows like the *Cattleyas* and flowers in the same way; the sepals and petals light purple, with a crimson lip. It blooms in October and November; continues in good condition three weeks. There are two varieties of this plant; one producing much darker coloured flowers, and being stronger in its growth than the other.

L. prestans.—A splendid dwarf evergreen species from Brazil; grows six inches high, and often blossoms twice a year; sepals and petals dark rose; lip rich purple; grows best on a block with a good supply of water at the roots in the growing season; of this there are several varieties, some of which are much better than others. This plant resembles *Cattleya marginata* in growth and size of flowers.

L. purpurata.—A magnificent Orchid from Brazil; grows two feet high and has light green foliage; blossoms so large and showy as to render this without exception one of the finest Orchids in cultivation. There are, however, many varieties of it, but all are well worth growing, though some are much finer than others; sepals and petals pure white, lip rich dark crimson-purple; in some of the varieties the sepals and petals are light rose; blooms during May, June, and July, and lasts three weeks in perfection, if the flowers are kept dry. Of this a fine specimen in bloom was exhibited at the Royal Horticultural Society's meeting last year by R. Warner, Esq., Broomfield; the plant was large and had about sixty flowers on it.

L. purpurata, var. Williamsii.—A splendid kind from Brazil, and certainly one of the finest of the genus for exhibition purposes; flowers large, three or four on a spike, and each bloom measures more than five inches across; sepals and petals beautiful delicate rose; lip rich crimson, and very large. Blooms in May and June, and continues three weeks in perfection; grows two feet high; foliage rich dark green.

L. Schilleriana.—A very fine species in the way of *elegans* as regards growth as well as flower; it grows eighteen inches high with light green foliage nine inches long; sepals and petals white; lip long and of a rich crimson: blooms during May and June, lasting three and four weeks in beauty. It makes a good exhibition plant on account of its showy flowers.

L. superbiens.—A magnificent Orchid from Guatemala. This is rather a large-growing plant, but it is one of the finest. The flowers are produced on a spike five feet long, having sometimes from fourteen to twenty flowers on one spike, each flower measuring nearly four inches across. The flowers are of a beautiful deep rose, variegated with dark red; the lip is a rich crimson, striped with yellow. It blooms during the winter months, and continues long in beauty. The finest plant I ever saw of this was in the Horticultural Gardens, Chiswick, and it is now in the collection of R. Warner, Esq., of Broomfield. This plant, when in bloom, is worth travelling miles to see. It sometimes produces as many as nine spikes of its beautiful flowers at one time; the plant is four or five feet in diameter; it is in perfect health.

L. xanthina.—Not a very showy kind, but worth growing, on account of its colour, which is pale yellow; flowers during May and June, and lasts in beauty three weeks.

LÆLIOPSIS.

Læliopsis Domingensis.—A pretty species, and the only one of the genus I have seen; it is evergreen, and compact in growth, with short bulbs; leaves about three inches long; the flower spike proceeds from the top of the bulb, and attains a height of twelve inches; blossoms rose-coloured, and produced at different times of the year, continuing in beauty for five weeks at a time; a very scarce plant, and best grown on a block suspended from the roof, with a good supply of water at the roots during the growing season; in growth this plant is like *Broughtonia sanguinea.*

LEPTOTES.

This small genus of Orchids deserves to be in every collection; they are small and compact-growing plants, with curious evergreen foliage in the shape of a thick rush, about three inches long, and produce their flowers from the top of the bulb. They are of easy culture, and will do either on blocks or in pots with peat. These plants are very accommodating, for they will grow in either house with a liberal supply of water in the growing season. They are propagated by dividing the plants.

Leptotes bicolor.—A pretty Orchid from Brazil; sepals and petals white, with a blotch of purple on the lip, and blooms during the winter months, lasting four weeks in beauty.

L. serrulata.—A charming little plant, from the same

country as *bicolor;* the flowers are nearly the same in colour; the only difference being, that they are much larger. It blooms in April and May; lasts three weeks in perfection. A scarce plant.

LIMATODES.

Limatodes rosea.—A charming winter flowering Orchid and the only species of the genus that I have seen; there are, however, many varieties of it. I have had five in flower this year, all of which differed more or less from one another in colours, which varied from white to dark rose and pink with different shades of rose. These few plants are more interesting or decorative during the dull season, and they may be had in flower for months together. All of them are deciduous, and have short thick bulbs which are nearly white, and flag-shaped leaves which proceed from the top of the bulb and attain a height of ten inches; during the growing season they require a liberal supply of water at the roots; in fact they should be kept well watered till they are in flower; after they have done blooming give them rest by withholding water; they do well in a cool house in pots with peat; pot them as you would *Calanthe vestita*, which they resemble in growth, and they flower about the same time as that equally useful Orchid; propagate by dividing the bulbs just as they begin to grow.

LYCASTE.

Of this genus several fine varieties have appeared since I last described it; these chiefly belong to *Skinneri*, and differ very much from each other in colour. I have seen eight in the collection of R. Warner, Esq., and also several fine varieties in the collection of J. Day, Esq. All the sorts of *Lycaste* have short thick bulbs and flag-shaped leaves; the flowers are produced from the side of the bulbs on spikes, about six inches in length; all the varieties are of easy culture, if properly attended to with respect to water, as they require a liberal supply during the growing season, especially *Skinneri* and its varieties. Mr. Gedney, gardener to W. Ellis, Esq., of Hoddesdon, managed these plants better than any cultivator with whom I am acquainted. In that collection, I have seen specimens with as many as fifty blossoms on them, producing a grand effect. Mr. Gedney kept his plants in a cool house, the temperature of which ranged from 50° to 60° in winter; he grows them in pots, in peat, with good drainage, watering liberally during the growing season; they should, in fact, never be allowed to get dry at the roots even while at rest. They are propagated by division after flowering.

Lycaste cruenta (from Guatemala).—Blossoms yellow, with a dark spot in the centre of the lip. Flowers in abundance in March and April, and lasts three weeks in perfection.

L. Deppii.—A good old species from Zalapa. Flowers during the winter and spring months, and lasts long in beauty.

L. Skinneri.—A beautiful free-flowering Orchid from Guatemala; sepals and petals pure white, tinged at the base with rose; lip same colour, spotted with crimson. Blooms during the winter months, lasting a long time in beauty. This plant ought to be in every collection; it is one of the finest for winter blooming, its large, numerous, singularly-formed, and richly-coloured flowers, rendering it at that season peculiarly attractive.

L. Skinneri alba.—A distinct and pretty variety from Guatemala; sepals and petals blush white; lip white; flowers smaller than those of *Skinneri*, with which it blooms contemporaneously, and it lasts a long time in perfection. A scarce variety.

L. Skinneri delicatissima.—Another distinct and handsome variety from Guatemala; flowers large, measuring six inches across; sepals and petals pinkish white; lip white, intermixed with rose. Blooms in February, and continues in perfection six weeks.

L. Skinneri rosea.—A magnificent variety also from Guatemala; flowers large, being quite seven inches in diameter; sepals and petals rich dark rose; lip white spotted, with crimson. This is the finest I have seen, and it lasts six weeks in beauty.

L. Skinneri superba.—Another splendid variety, likewise from Guatemala; sepals and petals blush white; lip of the richest crimson. A very showy kind.

MILTONIAS.

Of this class of plants there are some beautiful species, all of which are evergreen, compact-growing, and have light green foliage, with short bulbs, having two or three leaves on each; they flower freely from the side of the bulbs; some of them require treatment different from others, which shall be mentioned when describing the different species; most of them require to be grown in pots, in peat, with good drainage, and they like a liberal supply of water during the growing season, and the shadiest part of the house; they are propagated by dividing the bulbs when they begin to grow. They will succeed in the Mexican house.

Miltonia bicolor.—A beautiful species from Brazil, which grows about eight inches high in the way of *spectabilis*, but stronger and having larger flowers; sepals and petals white; lip also white, with a blotch of violet in the upper part. In bloom in August, and will last in perfection six weeks. Of this there are two varieties, one called *superba*, with larger flowers and more white on the lip.

M. candida.—A good species from Brazil; sepals and petals

yellow and brown; labellum pure white marked with pink. It produces its spikes of flowers during autumn, lasting three weeks in bloom.

M. candida grandiflora.—Much finer than *candida*; flowers larger and brighter in colour, and the plant is much stronger in its growth.

M. Clowesii major (from Brazil).—It produces long spikes of flowers in September and October; sepals and petals pale yellow barred with chocolate; lip purple and white; lasts long in perfection. Of this there are several varieties, but *major* is the best.

M. cuneata.—A pretty Brazilian species, which grows ten inches high, resembling *candida* in growth; the blossoms are produced on upright spikes, several together; sepals and petals dark brown, tipped with pale yellow; lip white. Flowers in February, and continues four or five weeks in perfection. It is best grown in a pot, in peat, with good drainage.

M. Karwinski.—A charming species from Brazil; sepals and petals pale yellow barred with brown; lip white spotted with chocolate. This fine species blooms during the winter months, lasting in beauty a long time.

M. Morelii.—A handsome Orchid from Rio Janeiro. The flowers of this species resemble those of *spectabilis*, the difference being that *Morelii* is much darker. Grows in the same way, and produces its flowers in September and October, continuing in bloom a long time.

M. Morelii atrorubens.—A magnificent variety of the preceding from Brazil, and grows in the same way; flowers large, often four inches across, and the colour much darker than that of *Morelii*. Blooms in September, and lasts a considerable time in beauty. A scarce plant.

M. Regnelli.—A charming and rare species, which grows in the way of *cuneata*, with light green foliage, twelve inches long; spikes upright, producing several flowers of a flat shape;

colour light rose and white. A very distinct kind, which lasts in bloom four or five weeks at a time. Generally blossoms in September or October, and is best grown in a pot, in peat. Also a rare Orchid.

M. spectabilis.—A beautiful Orchid from Brazil; grows about six inches high, and very free in producing its large solitary flowers in July and August, lasting six weeks in beauty if kept in a cool house and free from damp; sepals and petals white; the lip violet, edged with dull white. Of this beautiful species there are four varieties, some much better than others. A. Basset, Esq., of Stamford Hill, has some fine varieties of *spectabilis;* the lip of one which I saw in flower measured four inches across, and of a distinct and beautiful colour.

MORMODES.

There are several of this genus, but only three that I know of are worth growing; these are deciduous, and will do in either house, growing in pots, in peat, with a liberal quantity of water at the roots during their period of growth; afterwards they should be kept dry, and placed near the glass till they begin to grow. They are propagated by dividing the plant.

Mormodes citrinum.—A Mexican plant, and the best of the genus; flowers yellow, and produced on a short spike in July and August.

M. luxatum.—From the same country as the former; sepals and petals creamy white; lip the same, with stripes of brown in the centre. Blooms in July, lasting in bloom two weeks or more.

M. pardinum.—Also from Mexico; a beautiful species, with bright yellow flowers spotted with rich brown.

ODONTOGLOSSUMS.

To this magnificent class of Orchids so many fine additions have, of late years, been made, that it now contains some of the showiest in cultivation; all of them are evergreen; some have stout thick bulbs, very compact in growth, while others have small bulbs, with small narrow leaves; all produce their flower spikes from the sides of the bulbs, and they require an intermediate house to grow them in perfection; little heat is needed for them, many being destroyed by too much. I find them to succeed best in a cool house, the temperature of which ranges from 45° to 50°, with the exception of *O. citrosmum*, which does best along with the *Cattleyas;* some are best grown on blocks, while others do best in pots in peat with good drainage, a liberal supply of water at the roots being required during the growing season. They are propagated by dividing the bulbs just as they begin to push.

Odontoglossum citrosmum. — A charming Orchid from Guatemala. It produces its long pendulous spikes of flowers in June and July; the colour white, the flowers measuring two inches across, twelve or more on one spike. It will continue in perfection four or five weeks, if kept in a cool house free from damp, and makes one of the finest plants for exhibition.

O. citrosmum roseum.—A fine variety of the preceding, from Guatemala, grows in the same way, and produces its rosy flowers on long drooping spikes. J. A. Turner, Esq., exhibited this variety on one occasion, in fine condition, with many spikes, at the Manchester Botanic Gardens.

O. cordatum.—A desirable species from Guatemala; grows ten inches high, with short thick bulbs, from the side of which the spikes proceed; sepals and petals yellow barred with crimson; lip white, spotted round the edge with crimson. A rare plant, and one which requires to be grown in a pot, in peat.

O. coronarium.—A charming species from South America; grows eighteen inches high, with short thick bulbs and dark green foliage; the spike, which rises upright from the side of the bulb is about eighteen inches in height; sepals and petals reddish brown edged with yellow; lip bright yellow. It does best in a pot, in peat, and will continue a long time in perfection.

O. grande.—A remarkably handsome species from Guatemala; produces its very large flowers on an upright spike during the autumn. The flowers are mottled and striped with brown and yellow, like the back of a tiger; the lip white and purple. It remains in perfection three or four weeks, if the flowers are kept dry, and does best in a pot.

O. hastilabium.—A desirable Orchid from South America. It produces its spikes of flowers in June, July, and August, the colour being purple, green, and white. It continues blooming for two months, if kept in a cool house. This is a useful plant for exhibition, on account of its continuing in bloom so long.

O. Insleayii (from Mexico).—This grows in the same way as *grande;* the flowers are brown, yellow, and orange, and are produced on a short spike at different times of the year, lasting in beauty three weeks.

O. maculatum.—A good species from Guatemala; grows a foot high with short thick bulbs; produces its drooping spikes of blossoms during winter; colour yellow, crimson, and dark rose; will last in bloom a long time, and succeeds best in a cool house, in a pot, in peat.

O. membranaceum.— A beautiful small-growing species from Guatemala; sepals and petals pure white; the lip of the same colour barred with brown. It blooms during the winter months, and continues in bloom four weeks. This will do on a block of wood.

O. nævium.—A pretty dwarf Orchid from Truxillo, the colour of the flowers being white spotted with crimson; the lip yellow. It blooms in June and July, and lasts a considerable time in perfection. A scarce plant.

O. nævium majus.—A beautiful variety of the preceding from South America; grows ten inches high, and produces upright spikes of flowers, which are pure white speckled all over with rich crimson. Mr. Stone, gardener to J. Day, Esq., of Tottenham, has flowered some fine plants of this splendid variety, which is very rare; it requires great care to grow it well; a cool house suits it best, with a good supply of water; in short, it should never be allowed to become dry.

O. Pescatorei.—A magnificent species which grows ten inches high, which has small bulbs, and leaves a foot in length; produces branching spikes, richly ornamented with flowers, during April and May; sepals and petals white, with a shade of rose in them; lip white, yellow, and rose. This is also a difficult plant to manage, and requires great care. A. Basset, Esq., of Stamford Hill, has some fine plants of it; likewise J. Day, Esq., of Tottenham. There are several varieties of it, and all of them are good. It does best in a cool house in pots, in peat, with a good supply of water during the growing season.

O. Phalænopsis.—A magnificent species from South America; compact-growing, and having short bulbs and narrow leaves about eight inches long; flowers on a spike, generally two together, from the bottom of the bulbs, which are flat and very distinct from all other kinds; sepals and petals white; lip crimson in the centre, edged with white; will last in bloom four or five weeks. This is a difficult plant to cultivate, and very few grow it well; it requires a well-drained pot, good fibrous peat and a cool house. Mr. Byers, gardener to the Lord Chancellor of Ireland, has bloomed this several times.

O. pulchellum.—A good species from Guatemala; the flowers are white, with the exception of the crest of the lip, which is spotted with crimson. It blooms during the winter months, and lasts five weeks in good condition.

O. Rossii.—A desirable little Orchid from Mexico. It produces its white and purple flowers during the winter, and lasts long in beauty. Is best grown on a block.

O. triumphans.—A magnificent and rare species, which grows a foot high, with short thick bulbs, and produces branching spikes of blossom during the winter months; sepals and petals yellow, barred and spotted with crimson; lip white edged with dark rose; will last in perfection several weeks, and is best grown in a pot, in peat.

O. Uro Skinneri.—A pretty species from Guatemala; grows a foot high, with thick shiny bulbs; blossoms during autumn, and continues flowering for a long time; petals pale green spotted with brown; lip blush white; requires to be grown in a pot, in peat.

O. Warnerii.—A pretty dwarf species from Mexico; the flowers are yellow and crimson. It blooms in March and April, and lasts three weeks in perfection. A rare plant, and is best grown on blocks.

ONCIDIUMS.

A large class of Orchids, some of which are very beautiful; their flowers are rich and showy, and they make fine exhibition plants on account of their colours, which are very attractive in the midst of a collection of plants. They are all evergreen. Some of them are large growers, while others are more compact; some have handsome spotted foliage. They have generally short thick bulbs, from which the leaves and flower-spikes proceed; but in this respect there is much dissimilarity amongst them. Some kinds will do well on blocks of wood, but they are generally best grown in pots, with peat and moss and good drainage, with a liberal supply of heat and moisture in the growing season; afterwards only just enough water is required to keep their leaves and bulbs plump and firm. These plants are very accommodating, they will thrive in either house, and are propagated by division of the bulbs. The following are all fine sorts, and ought to be in every collection; they are of easy culture. There are many other *Oncidiums* worth growing besides those named in the following list.

Oncidium ampliatum majus.—One of the finest in cultivation from Guatemala. It produces its large yellow flowers in abundance, on a long branching spike three or four feet high, in April, May, and June, and continues blooming for two months when the spikes are strong. This is one of the finest *Oncidiums* we have for exhibition.

O. Barkerii.—A remarkably handsome dwarf Orchid from Mexico; the flowers are very large; the sepals and petals rich brown barred with yellow; the lip a bright yellow, about an inch and a half across. It produces its branching spikes of flower during the dull months of winter, which greatly enhances its value, and it lasts six weeks in bloom.

O. Batemanii.—A good distinct kind from Brazil, growing about eighteen inches high, and having pale green foliage; rather a shy-flowering plant, but one which is worth growing on account of its beautiful bright yellow flowers, which are produced at different times of the year.

O. bifolium.—A handsome dwarf-growing species from Monte Video: sepals and petals brown; lip bright yellow, and of large size. It produces short spikes of flowers in May and June, lasting a long time in perfection. There are two varieties of this plant; one much brighter in the colour of the flowers. This is best grown in a pot with moss, suspended from the roof, close to the rafters, with a piece of wire round the pot to hang it up by. This Oncidium is not so easy to grow as many others. Messrs. Loddiges, of Hackney, used to cultivate this plant better than any other grower of Orchids whose collections I have seen; they treated it in the way above recommended.

O. bicallosum.—A showy dwarf species from Guatemala; sepals and petals dark brown; lip bright yellow. This makes a fine plant for winter blooming, and continues in perfection a long time. It is very much like *Cavendishii* in flowers and growth.

O. bicolor (from the Spanish Main) is a fine species, which blooms in September. The lip is very large, of a deep yellow on the upper side, and almost white underneath; sepals and petals yellow, spotted with crimson. This will thrive on a block suspended from the roof.

O. Cavendishii.—A magnificent species from Guatemala. The large broad leaves of this princely Orchid are of a rich and lively green; its bright yellow flowers are produced in great abundance from strong and branching spikes, and, by appearing in the dull months of winter, greatly increase its value. The flowers, even at this season, retain all their brilliancy for several weeks. It may be grown on a block or in a pot, but on account of its size seems to do best in the latter.

O. ciliatum.—A pretty species from Brazil, and very compact in growth, being about six inches high; colour, beautiful brown and yellow; succeeds best on a block, but it must have good attention as to water at the roots.

O. crispum.—A good species from the Organ Mountains; a dwarf-growing plant with large flowers; sepals and petals a rich coppery colour; lip the same colour, with lighter spots in the centre. This generally blooms during the autumn, and lasts three or four weeks in beauty. Grows best on a block of wood.

O. crispum grandiflorum.—A very fine variety of the preceding. I only know of one plant of it, and that I bloomed, last year; it is now in the collection of J. A. Turner, Esq., of Manchester; its blossoms are very large—twice the size of those of *crispum*, and richer in colour. It succeeds best on a block, and requires a good supply of water at the roots.

O. divaricatum. — A small but abundantly-flowering species from Brazil; its yellow, orange, and brown-coloured flowers are produced on long branching spikes during the summer months: continues in perfection a long time. This is a useful plant for all purposes, when well grown: pot culture suits it best.

O. flexuosum. — A good old species from Brazil: producing showy flowers in abundance on a long spike; blossoms yellow, slightly spotted with brown: blooms at different times of the year, and continues blooming for several weeks. There are two varieties of this plant; one called *majus*, which has much larger flowers than the other, though of the same colour. This is a scarce variety, and is best grown in a pot with moss.

O. Forbesii. — A truly handsome dwarf species from Brazil: flowers large and very distinct, the colours being yellow, scarlet, and white: it blooms in November. This is a very rare plant; the only one I ever saw of it was in Messrs. Rollisson's collection at Tooting, where it was growing in a pot with peat.

O. hæmatochilum.—A fine species from New Granada. A compact-growing plant, in the same way as *Lanceanum;* sepals and petals greenish yellow blotched with chesnut; the lip a rich crimson and rose. A scarce Orchid.

O. incurvum. — A pretty distinct dwarf Orchid from Mexico: producing white and brown flowers during winter: lasts long in beauty, and is best grown in a pot with peat.

O. Lanceanum.—A remarkably handsome, distinct Orchid,

from Guayana, with beautiful spotted foliage ; flowers large and produced on a stiff spike about a foot or more high ; sepals and petals bright yellow blotched with crimson ; lip rich violet. Of this plant there are two varieties, one has the lip almost white : it blooms during the summer months, lasting four or five weeks in good condition, if the flowers are kept free from damp. This is best grown on a block or in a basket, with moss or peat, and makes a splendid plant for exhibition.

O. leucochilum. — A desirable and distinct species from Mexico, producing spikes, sometimes as much as ten feet long; sepals and petals yellowish green, lip a pure white : blooming at different times of the year, and lasting a long time in perfection : best grown in a pot.

O. longipes. — A beautiful compact small species from Rio Janeiro, growing about six inches high and producing spikes of flowers in great abundance ; lip large and of a bright golden yellow ; petals brown tipped with yellow : in blossom during the summer months, and will do well on a block, a very rare plant.

O. luridum guttatum.—A fine species from Jamaica : it produces long spikes of flowers, which are yellow, brown, and red in colour ; it blooms during the summer months, continues in perfection a long time, and is best grown in a pot with peat.

O. oblongatum. — A handsome free flowering species, from Guatemala : compact in habit with short thick bulbs ; foliage light green, about twelve inches in length ; flowers very showy, of a bright yellow and of good size ; blossoms during the winter months and continues a long time in perfection ; it succeeds best in a pot.

O. ornithorynchum. — A charming free flowering Orchid from Mexico ; grows ten inches high and produces graceful drooping spikes of flowers during the autumn and winter months ; colour, delicate rose and beautifully scented ; this does best in a basket, in which the flowers are shown off to good advantage; a general favourite with Orchid growers.

O. papilio majus.—A truly magnificent Orchid from Trinidad: with flowers the shape of a butterfly: it continues blooming from the old flower stems for years; as soon as one flower fades, another appears in the same place: the colour of the flowers is rich dark brown, barred with yellow; lip very large, with bright yellow in the centre, and edged with dark brown. This will do either in a pot or on a block. There are several varieties of *papilio*, but *majus* is the best.

O. phymatochilum.—A pretty species from Brazil, and very distinct from all others in flower as well as in growth; foliage dark green, attaining a height of twelve inches; bulbs thick; flowers very curious; sepals and petals yellow and reddish brown, lip white; this is a species which ought to be in every collection; it blossoms during May and June, and continues in flower two months at a time; it is best grown in a pot, in peat.

O. pulchellum.—A beautiful dwarf compact species from Jamaica, attaining a height of about six inches, with small bulbs and leaves, and it produces its spikes of white flowers in abundance during the summer months, remaining a long time in perfection; it thrives well on a block with plenty of moisture at the roots.

O. pulvinatum.—A free flowering Brazilian species compact in habit, and growing about a foot high; flower spikes not unfrequently ten feet long; colour of blossoms yellow, orange and brown; in bloom during the summer months, and will last long in perfection; pot culture and peat suits it best.

O. pulvinatum majus.—A fine Brazilian variety of the preceding, growing in the same way, but having blossoms much larger and brighter in colour; this I have only seen in the collection of E. McMorland, Esq., Haverstock Hill, who has flowered it very finely.

O. roseum.—A pretty small flowering species from Honduras: the flower is rose-coloured spotted with red: blooms at different times of the year; lasts six weeks in good condition, and is best grown in a pot with peat.

O. sarcodes.—A fine species, with handsome flowers, from Brazil: it produces its branching spikes of yellow and crimson flowers during March and April. This fine species was first flowered in the collection of Mr. Bunney, at the Stratford Nursery. A very rare, compact-growing plant; is best grown in a pot with peat.

O. sessile.—A pretty species, of compact growth, from Santa Martha: it produces its slender spikes of yellow flowers, spotted in the centre with pale cinnamon colour: blooms during the spring months. This fine species was flowered in 1850, by Mr. Iveson, then gardener to the Duke of Northumberland.

O. sphacelatum majus.—A good, free-flowering Orchid, from Honduras, producing its long branching spikes of flowers in April and May, lasting three or four weeks in beauty. The colour of the flower is yellow, barred with dark brown; it is best grown in a pot, with peat or moss.

O. unguiculatum (from Guatemala).—A pretty winter-flowering species: the large yellow flowers are produced on a long branching spike, three or four feet high, lasting a long time in perfection; pot-and-peat culture suits it best.

O. variegatum.—A pretty species from the West Indies; grows six inches high and has dark evergreen foliage; spikes branching and upright, bearing many blossoms of a rose and pink colour, and continuing in beauty several weeks; it is best grown on a block with plenty of moisture at the roots.

PAPHINIAS.

Of this useful genus there are only two with which I am acquainted, but both of them are well worth growing; they are very compact in habit, having short bulbs and flag-shaped leaves from six inches to a foot in length; they succeed best in the Cattleya house in pots in good fibrous peat with plenty of drainage, liberally supplying them with water at the roots during the growing season; propagation is effected by dividing the bulbs.

Paphinia cristata.—A pretty dwarf free flowering species from Demerara; grows eight inches high, and has short shiny bulbs, from the base of which the flowers appear on a short stalk, three together, and droop downwards over the side of the pot; sepals and petals dark chocolate tinged with purple; lip white, barred with purple and fringed; of this there are two varieties, one with much darker flowers than the other. In bloom at different times of the year, and will remain in beauty about two weeks.

P. tigrina.—A fine species from Trinidad; grows ten or twelve inches high, and produces its flowers, which will last several weeks in perfection, on an upright spike, several together; a scarce plant, and one which is best grown in a pot, in peat, with good drainage.

PERISTERIAS.

A singular class of Orchids, of which there are several species, but only three that I can recommend as worth growing. There is one noble plant in this class, *Peristeria elata*, the dove plant, which ought to be in every collection. These plants furnish flowers from the bottom of their large bulbs, and will grow in either house: they are best grown in pots, with loam and leaf-mould, with a good quantity of water during their period of growth; afterwards give them a good season of rest, and keep them nearly dry at the roots: if allowed to get wet during their rest they are apt to rot. They are propagated by dividing the plants.

Peristeria elata. — A noble free growing Orchid from Panama: with leaves three or four feet high, rising from large bulbs five inches high, and sending up its tall spikes of white, waxy flowers, in July, August, and September: the central parts of each flower present very striking resemblance to the figure of a dove. This plant continues blooming for two months when the spikes are strong.

P. cerina (from the Spanish Main): it produces bunches of yellow flowers close to the bulbs, and blooms in June or July.

P. guttata (from South America).—A curious Orchid, which produces bunches of flowers close to the bulbs, whence they hang over the edge of the pot: colours red and yellow. It blooms in September, lasting two or three weeks in perfection.

PHAJUS.

A fine class of terrestrial Orchids, very free in producing their beautiful spikes of flowers, and when well grown are noble objects. They are of easy culture, and will repay attention and care. They are large-growing plants, with noble foliage: of this class there are not more than three, that I know of, that are distinct and worth growing. All need the same treatment, excepting *albus*. They all require plenty of heat and moisture at their roots in their growing season, but the water should not touch the young growth. *Phajus Wallichii* and *grandifolius* are best grown in pots of large size, with loam, leaf-mould, and rotten cow-dung, as directed in the remarks on Terrestrial Orchids. They are propagated by dividing the bulbs after they have done blooming.

Phajus albus.—This stately Indian plant flowers in July and August. The blossoms are pure white, except the lip, which is pencilled with purple : continues blooming for five weeks. It is a deciduous species, loses its leaves after the growth is finished, and is best grown in a pot with rough, fibrous peat, and good drainage, with a liberal supply of water at the roots in the growing season. After the growth is completed it requires a good rest, by being placed in the cool house, and kept dry, till it begins to grow, when it should be put into heat, and treated as before. See Hints on Propagation respecting this plant, of which there is a variety with flowers wholly white.

P. grandifolius. — This noble evergreen plant comes from China, grows three feet high, and produces its flower-spikes one or two feet above the foliage, the colour of the flowers being white and brown ; and blooms during the winter and spring months, and lasts long in beauty if in a cool house. This is a most useful plant for winter blooming, and a noble plant for exhibition.

P. Wallichii.—An Indian Orchid, and one of the finest in cultivation : a large-growing plant, about four or five feet high, producing long upright spikes of flowers in March, April, and May. The blossoms are orange-yellow in colour, or buff tinged with purple : it will keep blooming for six weeks. This makes a noble plant for exhibition. For further notice of these plants, see remarks on Preparing Orchids for Travelling.

PHALÆNOPSIS.

An exceedingly fine genus, containing comparatively but few species of not very large growth, but producing magnificent flowers, which last long in beauty. Within the last month or so has been added to it one of the finest of all Orchids, viz., *P. Schilleriana*, which has really beautifully variegated foliage, as well as wonderfully distinct and fine flowers; other kinds are also all compact handsome plants, with beautiful thick fleshy leaves, from whose axils are produced spikes of charming flowers. As regards growth they require very little room, and they may be had in bloom all the year round. I have seen *grandiflora* blooming for six months; I have exhibited the same plant for seven years at six exhibitions each year, and sometimes with as many as from seventy to eighty flowers expanded on it at one time. This same plant, I believe, is still in the collection of C. B. Warner, Esq., at Stratford. I mention this fact to show how long such a fine plant may be had in perfection; for, if I recollect rightly, it is quite fourteen years since we first exhibited it at Chiswick and Regent's Park.

There are now five sorts of *Phalænopsis* in cultivation in this country, and all are well worth growing; they have all beautiful flowers and handsome evergreen foliage; they are compact in their habits of growth, free flowering, and, as I have just pointed out, continue a long time in perfection— all qualities proclaiming them to be plants of more than ordinary value. They all require the same kind of treatment, and an East Indian heat, together with a good supply of moisture during their growing season. Plants of this genus are found in Java, Manilla, and Borneo, where the heat is high, and of course ought to be imitated under artificial circumstances as near as possible; they are found growing on the branches of trees in damp, moist places.

Nevertheless, they are of easy culture, and if properly attended to are seldom out of order; unlike many other Orchids they have no thick fleshy bulbs to support them, and of course to have them in perfection they require more nourishment. This is done by giving them more moisture at their roots during the growing season; in fact they should never be allowed to get dry; if so they are apt to shrivel and often lose their bottom leaves, which spoils their appearance, for the beauty of the plants consists in their having good foliage as well as good flowers. The growing season is from March to the end of October, during which time the temperature by day should range from 70° to 75°, allowing it to rise to 80° or more by sun heat provided the house be shaded. The night temperature should range from 65° to 70° in March and April. Afterwards it may be allowed to rise a few degrees higher. During their resting season, from the end of October to February, the temperature should range from 60° to 65° by night, and 65° by day, or even a little more will not do any harm with sun heat. In giving air, a little should be admitted close to the hot-water pipes, so that it may be warmed on entering the house; a little moisture should be sprinkled about on fine days, but let it be in the morning so that the house may be dry by night.

Phalænopses are grown in different ways; some are placed on blocks, some in pots, and others in baskets. I have found them to succeed well under all three modes of treatment, but they require more moisture at the roots if grown on blocks; if in pots, give more drainage than in baskets. The best way is to place an empty pot upside down in the bottom of the one you intend for the plant, and then fill in with potsherds broken into pieces about two inches square to within two inches of the rim; then fill up with sphagnum, having a few small potsherds mixed with it, and elevate the plant three inches above the rim, taking care to keep it well above the moss. The successful culture of these plants, as well as of all others, depends upon efficient drainage. If grown on blocks they should be placed on a good-sized one, so that there is room for the roots to cling to it. In fastening the plant on, first place a little live sphagnum on the block, and afterwards fix the plant on with copper wire and hang it up to the roof, but not too near the glass, or your plants may get injured by cold, which should be guarded against during winter. Many Orchids are injured in that way. If they should get into an unhealthy condition, the best plan is to turn them out of their pots or baskets, and shake all material off their roots, wash them with clean water, cut off all the decayed parts, and replace them on blocks with a little sphagnum, giving them a good supply of moisture; place them at the warmest end of the house, with not too much light; under this treatment they will soon begin to root and improve in appearance. I need hardly add that they should be kept clear of insects, especially thrips, which soon disfigure the foliage; it is best kept under by constantly washing the leaves with a sponge and clean water, or fumigating the house with tobacco smoke,—an operation which should be conducted with great care.

Phalænopses are difficult to propagate; sometimes they will produce young plants on the old flower stems, which should be left on till well rooted; then place them on small blocks.

Phalænopsis amabilis.—The queen of Orchids. This magnificent plant comes from Manilla. It produces its graceful spikes of flowers nearly all the year round: the flowers, each of which is three inches across, are arranged in two rows down the spike; sepals and petals pure white; the lip of the same colour, the inside streaked with rose-pink. The flowers continue in perfection a long time, if they are kept free from damp; if the flowers get wet they are apt to spot.

P. grandiflora.—A truly handsome variety from Java. The flowers are produced in the same way as in *amabilis*. The only difference between the two is, that *grandiflora* has much larger flowers, with yellow in the centre of the lip, instead of pink; and the leaves are longer, and of a lighter green. This makes a fine plant for exhibition.

P. Lobbii.—A pretty species, and one which is very

rare; it grows in the same way as other kinds, with pale green foliage. Of this there was only one plant until within these last few months, during which I have imported some with other kinds, now in my collection.

P. rosea (from Manilla).—A small flowering species, but very pretty; grows in the same way as the two former, but very inferior in beauty; the colour being white, slightly tinged with pink; the lip deep violet. It blooms at different times of the year, lasting long in beauty. A scarce Orchid.

P. Schilleriana. — A magnificent new species from Manilla, which I had the good fortune first to introduce to public notice, and one which, I have no doubt, will prove the finest in cultivation; it has, as I have said, beautifully variegated foliage, and is quite distinct from all other kinds; the leaves are similar in form and equal in size to those of *P. grandiflora*; ground colour dark green, interspersed with irregular bands of white; the flower spikes are produced from the axils of the leaves. In their native country they are more than three feet long, and more branched than those of the other kinds. I have a dried spike, on which there has been more than one hundred blossoms. On the plant I have in bloom each individual blossom measures more than two and a half inches across, and they are arranged in two rows down the spike; sepals and petals beautiful light mauve edged with white; lip the same colour, with darker spots, with the exception of the upper part, which is yellow spotted with reddish brown; the inside is handsomely spotted, indeed the whole aspect of the plant is very attractive. Several plants of this have flowered within the last few weeks, and I find that they differ from one another in shade of colour. R. Warner, Esq., of Broomfield, exhibited a fine plant of this before the Floral Committee of the Royal Horticultural Society. E. McMoreland, Esq., Haverstock Hill, also exhibited one with a little more colour in the flowers than Mr. Warner's specimen. Of this plant the roots are very distinct from those of others of the genus; they are flat, and have a rough appearance, and very free in growth; the blossoms continue several weeks in perfection. Mr. Warner has had it in bloom seven weeks, and it is still in fine condition; a great recommendation. The flowers are sweet-scented.

PLEOINES.

Pretty small growing distinct looking plants, with small bulbs. Every year after finishing their growth, the leaves naturally begin to die away, which is the proper time to rest them, when they should not have much water at the roots— only just enough to keep them from shrivelling. They begin to flower when they begin to grow. In their native country they are called Indian crocuses, throwing up, as they do, their flowers much like our common crocus, except that they are of a different shape; the blossoms are very handsome and rich in colour; they are produced on stems three inches high, in autumn, when destitute of leaves. Though not often seen, these plants are well worth a place in every collection. Being deciduous, they get neglected and lost just at a time when they should receive most attention. They give little trouble if properly managed, and at the right time, which is the great secret in the treatment of most deciduous plants; they require a good season of growth, and after that one of rest. The way in which I treat them is to pot them as soon as they have flowered in a mixture of loam, peat, and sand, with good drainage, and plenty of water; while growing, after the bulbs are formed, give them but little water—just enough, as I have stated, to keep them from shrivelling. As soon as they begin to show flower, water freely, which will induce their blossoms to come finer; they require the heat of a Cattleya house.

P. humilis.—A splendid species from India, dwarf-growing, in the way of *Wallichiana*, and with flowers three inches in diameter; sepals and petals bright rose; lip white, spotted with crimson, and striped with brown. Lasts in beauty two or three weeks together.

P. lagenaria.—A fine dwarf species, also from India, in the way of *maculata;* flowers on single stems three inches long; blossoms three inches across; sepals and petals mauve; lip white, veined with crimson.

P. maculata.—A beatiful Orchid, from the Khoosea Hills, with leaves six inches long; sepals and petals delicate white; lip same colour, beautifully barred with crimson. It blooms in October and November, continuing three or four weeks in perfection.

P. Wallichiana.—A truly handsome dwarf species from India, producing its solitary flowers in October and November, and lasting two weeks in beauty. The colour is of a deep rose, and the lip the same, with a dash of white in the centre; the blossoms are three or four inches across.

PROMENÆAS.

This is a small genus of pretty, little, dwarf-growing Orchids, about three inches high; they produce their flowers from the side of their bulbs, and hang over the edge of the pot. They are not very showy plants, but curious, and ought to be in every collection, and are best grown in pots with peat, in either house, with the same treatment as *Paphinia cristata.*

Promencea Rollissonii (from Brazil).—It produces its pale yellow flowers during the autumn, lasting three weeks or more in beauty.

P. stapelioides (also from Brazil).—The colours of the flowers are green and yellow. It blooms in July, August, and September, and lasts long in perfection.

RENANTHERA.

Renanthera coccinea.—A Chinese Orchid, and the only one of the genus that I have seen worth growing. It is a straggling plant, often twelve feet or more in height, with long stems, furnished with leaves up each side of them; the latter are about three inches long; flowers beautiful orange scarlet, produced from the axils of the leaves, and they continue in bloom several weeks together. This plant does best trained up in the house where it is well exposed to sunlight, which is the only sure way of inducing it to bloom; it is indeed a shy bloomer, which doubtless accounts for it not being much cultivated. By letting it grow up the roof, where it gets plenty of light and sun, it will, however, most likely flower when sufficiently strong for that purpose. I have seen it do well on a large block suspended from the roof, where it had as many as five spikes on it at one time, and when well bloomed it is well worth all the care that is bestowed upon it, for it is a really handsome plant. As regards treatment, grow it well during summer, and rest it in winter, giving a liberal supply of heat and moisture during the time it is in vigorous growth, which is from March till October. While resting, let it have but little water—just enough to keep it from shrivelling. The most suitable material for it is sphagnum moss at the bottom of the plant, which should be kept moist during the growing season.

SACCOLABIUMS.

Some of the finest Orchids in cultivation belong to this class. They are very compact in their growth, with beautiful, long, and pendent evergreen foliage. Their habit of growth is the same as that of the *Aerides*, and they require the same heat and treatment. These plants inhabit the hottest parts of India, and are found growing on the branches of trees. They produce their long, graceful racemes of flowers, which are often a foot and a half long, from the axils of the leaves. They are propagated in the same way as the *Aerides*, and are infested by the same sorts of insects. The following list comprises the best among this beautiful class of plants. There are several others; some of which are not worth growing. I have only named those that are the most beautiful, and which ought to be in every collection, however small: they are even handsome without flowers.

Saccolabium ampullaceum.—A distinct compact habited and pretty Orchid from India, with leaves about three or four inches in length; grows about ten inches high; producing spikes of flower about six inches long, in May and June, of a beautiful violet colour. Will succeed on a block, and remains in beauty three weeks.

S. Blumei.—A beautiful distinct species from Java. It produces its flowers in July and August, which are violet and white in colour, and last three weeks in perfection.

S. Blumei majus.—A charming Orchid from Java. The colours of the flowers are the same as *Blumei*, the difference between the two consisting in the *majus* having much larger flowers and finer spikes. The growth of the plant is also much stronger than *Blumei*.

S. curvifolium.—A handsome, compact habited dwarf-growing species, also from India; height six inches to a foot; foliage light coloured; a free flowering kind with orange scarlet blossoms. Will thrive well on a block suspended from the roof; blooms in May and June, and till within this last twelve months, very rare, but now more plentiful.

S. furcatum.—A distinct and fine species from India; grows somewhat slowly, and has short leaves about eight inches long; the flowers of this are further apart on the spike than in *guttatum;* colour white, spotted with rose; in bloom during July and August, and continues in perfection three weeks.

S. guttatum.—A remarkably handsome species from India. It blossoms from May to August. The flowers are white, spotted with deep rosy-purple. It remains three or four weeks in perfection, if removed to a cooler house, and kept free from damp. There are two or three varieties of this plant. This makes one of the finest plants for exhibition. I observed specimens of this plant shown in the year 1850, with as many as twenty or twenty-five spikes of flowers on a plant at one time.

S. guttatum giganteum.—A magnificent variety, the leaves of which are longer than those of *S. guttatum*, and spotted; spikes also longer, and the flowers more distinct in the markings than in the former; makes a fine exhibition plant; blossoms in June and July, and will last three or four weeks in perfection.

S. guttatum Holfordianum.—Another variety and the finest of the genus; leaves broader than in *S. guttatum*, and more blunt at the ends; flower spikes much larger, longer, and of a richer colour than those of the kind just named. This was bloomed first by Mr. Bassett, gardener to R. S. Holford, Esq., Weston Bert, Gloucestershire, in compliment to whom it is named. Of this there is also a fine plant in the collection of A. Bassett, Esq., Stamford Hill, under the care of Mr. Baker, his gardener.

S. miniatum.—A pretty, distinct, small-growing Orchid from Java. It is not so good as the other species named, but it is worth growing. It produces its short spikes of vermilion-coloured flowers in March and April, lasting three weeks in beauty. This will do well on a block of wood without moss.

S. præmorsum.—A lovely species from Malabar. The flowers are white, thinly spotted with lilac. It blooms in May and June: lasts three weeks in perfection. A slow growing kind, with stout stiff foliage. It also makes a fine exhibition plant.

S. retusum.—A fine free-growing species, stronger than most others; blossoms in May and June, producing long spikes of white and pink-spotted flowers in great abundance, and continuing in bloom three or four weeks. A useful plant, which comes in rather earlier than any of the other kinds.

S. Wightianum.—A small and pretty species from India, not so good as any of the others that I have named; but still worth growing, on account of its compact habit and distinct coloured flowers, whose sepals and petals are orange-yellow; lip violet; blossoms in June, and will last three weeks in perfection. Thrives well on a block.

SCHOMBURGKIAS.

Of this genus only a few are worth cultivating, though the blossoms of many are individually very attractive; the fact is, they are too shy in flowering, producing long spikes with but very few blossoms on them. In growth they resemble *Cattleyas* or *Lælias*, except that they are not so compact, producing upright bulbs twelve inches or more in length, with generally from one to three leaves on the top; their spikes proceed from the top of the bulbs, and are from three to five feet long. The following are the best I have seen; they succeed on blocks or in baskets suspended from the roof in moss and peat mixed together in the Cattleya house; they like a liberal supply of water during the growing season; after they have completed their growth keep them dry until they begin to show flower. Propagation is effected by dividing the bulbs.

Schomburgkia crispa.—A pretty species, from Brazil, with bulbs a foot high, and spikes three feet long, with several flowers on the top. This does best in a basket, and will grow to a good size; blossoms during winter, and will last three weeks in perfection, if the flowers are kept dry.

S. Lyonsii.—A handsome kind, which grows a foot high,

and produces its spikes of blossoms three or four feet long, several together; colour of flowers white, spotted with purple; will succeed either on a block or in a basket suspended from the roof, and lasts in bloom three weeks.

S. tibicina.—A species from Honduras, and the best I have seen of the genus; bulbs, large when well grown, and will not flower till the bulbs are strong; the blossoms are produced on long spikes five feet or more in length, many together; each flower measures more than two inches across; sepals and petals pink, spotted with rich chocolate; lip white with rose colour on the side; blooms in May and June, and will keep flowering for six weeks; it requires to be grown strong before it will blossom, and is best grown on a block with plenty of heat and moisture during the growing season.

SCUTICARIAS.

A small genus, of which I know of only two species, both of which have evergreen foliage in the shape of a rush. The flowers, which are handsome, proceed from the bottom of the bulb, on a short stalk about two inches high. They are best grown in the East India house, either on blocks or in baskets with moss, with a liberal quantity of water at the roots in the growing season. They are propagated by dividing the plants just as they begin to grow.

Scuticaria Hadwenii.—A pretty Orchid with foliage a foot long. This plant is very much like *Steelii;* the only difference between the two is, that *Hadwenii* furnishes flowers more erect, and stands one or two inches higher. The colour of the flowers is nearly alike. A rare plant.

S. Steelii.—A handsome plant from Guayana, the foliage of which is pendulous, three or four feet long; flowers yellow, spotted with crimson. It produces its blossoms at different times of the year, lasting a good time in perfection. Of this plant I saw three varieties growing in the collection of — Brocklehurst, Esq., under the care of Mr. Pass, his gardener.

SOBRALIAS.

There are some handsome varieties belonging to this family. Their colours are brilliant, and flowers large. They are strong, free-growing plants, with evergreen foliage, and produce their flowers out of a spathe, one at a time, from the top of the reedy-like stem : they have as many as six flowers from each spathe; as soon as one decays another appears. They will grow either in the East India house or the Mexican, and thrive best in pots of a large size, with rough, fibrous peat, and about three inches of drainage, with plenty of water at the roots in the growing season; afterwards keep them much drier. When these plants get too large, turn them out of the pot, and part them, making two or three young plants, which will soon grow and make flowering plants. The following are the best of this class. There are several others, but they seldom compensate for the room and care they require.

Sobralia liliastrum.—A distinct species from Guiana; the flowers are white, and are produced in July and August, and last but a short time in beauty. There are two varieties of this, one much better than the other.

S. macrantha.—A remarkably handsome Orchid from Guatemala; the finest of the genus. The flowers are large, six inches across, of a beautiful rich purple and crimson. Blooms during the summer months, and lasts three days in perfection. This makes a fine plant for exhibition.

S. macrantha splendens.—A charming variety from Guatemala, flowering from June to August. It has darker flowers than *macrantha*, but not so large.

SOPHRONITIS.

Another small genus of Orchids, and one of them, *grandiflora*, very handsome. The others are worth growing, on account of their not taking up much room; they are small, and are best grown on blocks of wood, with a liberal supply of heat and moisture all the year. They are evergreen, and produce their flowers from the top of the bulb, and are propagated by dividing the plants just as they begin to grow.

Sophronitis cernua.—A small-flowering Orchid from Rio, with small bunches of red flowers, several together. It blooms during the winter, lasting long in beauty.

S. grandiflora.—A truly handsome species from the Organ Mountains; the flowers are large, of a beautiful bright scarlet colour, lasting six weeks or more in perfection. It blooms in November and December. This plant ought to be in every collection. Of this species there are two varieties; one produces short obtuse leaves, with exceedingly brilliant-coloured flowers; the other has longer leaves and bulbs, but fainter-coloured flowers.

S. violacea.—A pretty, distinct species, from the same country as the former. It produces its violet-coloured flowers during the winter months, and lasts long in beauty.

STANHOPEAS.

Fine looking Orchids, most of which have handsome-coloured, waxy-looking, and curiously-formed flowers, produced from the bottom of the plant on spikes, several together. The flowers only last a few days. Bulbs short with strong evergreen foliage, about a foot and a half high. *Stanhopeas* are of easy culture, and are best grown with moss in baskets suspended from the roof; they require a liberal supply of water in the growing season. After their growth is completed they are the better for a good period of rest, during which they should be kept nearly dry at the roots. They will grow in either house. The baskets should be made shallow, and open at the bottom and sides, so that the flowers may easily find their way through. I shall only name a few of the best sorts, though there are others worth growing where room is not an object.

Stanhopea aurea (from Guatemala).—The colour of the flowers is yellow. It blooms during the summer and autumn months.

S. Devoniensis (from Mexico).—The flowers are orange, spotted with red. It blooms during the summer.

S. grandiflora.—A distinct species from Trinidad. The flowers are white, and very large. It blooms at different times of the year, and lasts but three days in flower.

S. insignis (from Trinidad).—The colour is pale yellow, spotted with red; the lip nearly white spotted with purple. It produces its flowers in August and September.

S. Martiana (from Mexico).—The sepals and petals are straw-coloured, spotted with red; the labellum white. Blooms during the autumn.

S. oculata (from Mexico).—Flowers from July to November; the colour of the flowers is pale yellow, spotted with purple.

S. tigrina (from Mexico).—The handsomest of the genus; the blossoms are very large, a pale yellow, barred and blotched with deep chocolate. It blooms in July, August, and September.

S. tigrina lutescens.—This is also a fine variety from Guatemala; the flowers are brilliant yellow, inclining to orange, and barred with deep chocolate. Blooms at the same time as *tigrina*. This makes a fine plant for exhibition, on account of its large showy flowers.

TRICHOPILIA.

This is a small genus of Orchids, some of which are very handsome and distinct; their flowers are very curious in form, and are produced from the side of the bulbs. They are dwarf evergreen plants, and are best grown in pots, with peat and good drainage, and should be well elevated above the rim of the pot on account of their drooping flowers, without too much water at the roots at any time. They will do best in the Mexican house, and they are propagated by dividing the plant.

Trichopilia coccinea.—A pretty species from Central America. A very distinct plant; sepals and petals are twisted, brownish, with yellow; the lip of a deep crimson, with a narrow edge of white. It produces its blossoms in May and June, lasting three weeks in beauty. Of this plant there are many varieties.

T. crispa.—A charming species, and very rare; it resembles *coccinea* in growth, except that the bulbs are larger. Produces its blossoms during May and June, and they last two weeks in perfection. Requires to be grown in a pot, and elevated three inches above the rim, to show off to better advantage the flowers which are drooping. Makes a good plant for exhibition. Very scarce.

T. picta.—A beautiful new species, which grows six inches high; bulbs about two inches long. A very distinct kind; the flowers of which are of a pale yellow, and spotted with brown. In bloom in August and September, and continues two weeks in good condition.

T. suavis.—A magnificent species, and very handsome; grows in the way of *Odontoglossum grande*; the bulbs and leaves greatly resemble that plant. The flowers are white, spotted with pink, and as many as three are produced on a spike. It blooms in March or April, lasting about two weeks in perfection, and is best grown at the coolest end of the house.

T. tortilis.—A fine species from Mexico. The sepals and petals are twisted like a corkscrew ; they are brown and pale yellow ; the lip white, spotted with red. It produces its flowers freely at different times of the year, and lasts two or three weeks in beauty. There are two varieties of this plant, one with much brighter coloured flowers than the other.

UROPEDIUM.

Uropedium Lindenii.—A singular Orchid, and the only one of the genus that I have seen ; it is a compact growing plant, with pale green foliage about ten inches in length. In growth it resembles *Cypripedium caudatum,* and throws up its flower spikes from the centre of the leaves ; the blossoms, which are produced two or more together, are large and singular in shape ; from the end of the petals are produced three long tails, measuring from eighteen inches to two feet in length, of a reddish brown colour, the flowers themselves being greenish white, striped with dark green. Though not very showy, this plant is well worth growing on account of its peculiarly-shaped blossoms ; it is by no means a difficult plant to cultivate, if it but gets the treatment it requires. I have found it to do best in a pot in a mixture of loam, peat, and sand, with good drainage ; place the plant just below the rim of the pot, and water liberally at the roots during the growing season, which is nearly all the year ; it requires but little rest ; having no fleshy bulbs to support it, it of course wants a certain degree of moisture at the roots, even when comparatively at rest ; after growth is completed it will begin to show flower, and then care should be taken that it does not get dry at the root, for a certain quantity of nourishment it must have to help it to bring its flowers to perfection. I have seen this plant shrivelled when the blossoms have been showing, and through that the latter have been nearly spoiled and the plant injured; for if allowed to get into an unhealthy state it is a long time before it recovers; it blooms during the summer months, and lasts some time in beauty; propagation is effected by dividing the plant when done growing, or just as it begins to push.

VANDAS.

These are a lovely class of plants, with magnificently-coloured flowers, some of them very large. There are not many plants that surpass *Vandas* in beauty of flowers. They grow in the same way as *Aerides* and *Saccolabiums*, having gracefully-formed evergreen foliage; the upright spikes, all of which bear large waxy flowers, spring from the axils of the leaves. They require treatment similar to that of *Aerides*, being subject to the same sort of insects, and are propagated in the same way.

Vanda Batemani.—A noble Orchid. A large upright-growing plant, which blooms in July, August, and September, and continues blooming for three months. The colour of the flowers is yellow, spotted with crimson, the back of the sepals and petals being rose-colour. A very rare plant.

V. cœrulea.—A remarkably handsome Orchid from India. This fine plant produces its upright spikes of flowers, nine or ten together, five inches across. The colour of the flowers is a rich lilac; they are produced during the autumn months, and last six weeks in perfection. This plant does not require so much heat as the other kinds.

V. cristata.—A charming Orchid from India; sepals and petals are whitish; the lip is spotted and striped with dark brown; it produces its flowers from March to July, lasting in bloom six weeks or two months. This very scarce Orchid makes a fine plant for exhibition.

V. gigantea.—A magnificent Indian plant, with noble dark green foliage, three inches broad, and of graceful habit; if the flowers are at all equal to the foliage it will indeed be one of the finest of its class. I have not yet seen it in bloom, but I believe it to be a very fine kind; it is reputed to produce a drooping spike, with many flowers, three inches in diameter, of a deep yellow, with crimson spots. I have a fine specimen showing flower, which I hope to see in blossom this season.

V. Lowii.—A handsome-growing plant from Borneo, with dark green foliage, grows several feet high, and is very shy in producing its long drooping spikes of flowers, many together, which are reddish brown and buff colour, two of which, at the base of the spike, are yellow; it blossoms during the summer months; of this there are several large plants, but I only know of two that have flowered, and these were in Messrs. Veitch's collection.

V. insignis.—A pretty free-flowering species from Java, and one which makes a fine specimen, being free in growth; petals pale-yellow, spotted with crimson; lip pale lilac; blooms at different times of the year, but generally in spring and autumn, and continues in bloom six weeks.

V. Roxburghii.—A good old species from India, with white and purple-coloured flowers, which appear during the summer, and last five or six weeks in beauty. There are two varieties of this plant; one is much handsomer and has a darker coloured lip than the other.

V. suavis.—A truly magnificent Orchid from Java; a strong-growing species, and very free in flowering. It produces branching spikes of flowers, each being large, of a creamy white, spotted with crimson. It blooms at different times of the year, lasting a long time in perfection. This makes one of the finest plants for exhibition.

V. teres.—A handsome, curiously-growing, and distinct Orchid, the foliage resembling a rush. It comes from Sylhet. Its large red and yellow-coloured flowers are produced from June to August, and last four or five weeks in beauty. It is rather a shy flowering species, and is best kept rather dry during the winter, to make it flower. It thrives best on a block of wood, the block being plunged into a pot.

V. tricolor.—A charming free-growing species from Java; grows in the same way as *suavis;* the sepals are pale yellow, spotted with crimson; lip purple, striped with white. It blooms at different times of the year, and lasts long in perfection. There are two or three varieties of this plant; some are not so good as others. This also makes a fine plant for exhibition.

V. tricolor superba.—A charming Orchid from Java, and a fine variety of the preceding, often called *suavis* (Rollisson's). Of this there are many varieties, the best of which is nearly equal to *suavis* (Veitch's).

V. violacea.—A magnificent species from Manilla, which grows in the way of *Saccolabium Blumei majus*, with stiff pale green leaves, enriched with several distinctly-coloured lines running down each leaf; petals white, with violet spot; lip violet, beautifully marked with a darker colour; the flower spikes are produced in the same way as those of *Saccolabiums* in a drooping manner, very distinct from all the other *Vandas*, and it is more compact in growth. Till within the last few months this plant has been very rare, but now I have many plants of it in fine condition.

WARRÆA.

Warræa cyanea.—A very pretty and distinct Orchid from Columbia. It is an upright-growing plant with evergreen foliage, and requires to be grown in a pot, with peat and good drainage, in the East India house. The colour of the flowers is white and purple. It blooms in June, lasting a long time in beauty. A rare species. There are several more of this tribe, but I have not seen them in bloom.

W. tricolor.—A very good species from Brazil. The flowers are produced on an upright spike, two feet high, in June and July; sepals and petals white; the lip white, with yellow and purple in the centre: the blossoms last a long time in perfection, and require the same treatment as the former one.

ZYGOPETALUMS.

Handsome plants with evergreen foliage. They generally bloom during the winter, which makes them very valuable. They are rather large-growing plants, of easy culture, and will do in either house in pots, with peat and good drainage, and plenty of water at the roots. They are propagated by dividing the plants.

Zygopetalum brachypetalum.—A Brazilian species and one of the handsomest of the genus, having brownish sepals and petals; a little marbled with green and a deep blush violet; lip veined with white. It blooms in December, lasting long in perfection.

Z. crinitum cœruleum.—A handsome variety from Brazil. It produces upright spikes, sometimes two from the same bulb, and bears beautifully variegated flowers during the winter. The sepals and petals are green, barred with brown; the lip is white or cream-coloured, streaked with bright blue.

Z. intermedium (also from Brazil).—It produces green and blue blossoms during the autumn, and continues in perfection four or five weeks.

Z. Mackayi.—A handsome Brazilian plant. It produces long spikes of large flowers during the winter. The colour or the flowers is greenish yellow, spotted with brown and lilac.

There are several varieties of this plant; some much finer than others. It lasts in perfection a long time.

Z. maxillare.—A free-flowering, pretty species from Brazil, producing its drooping spikes at different times of the year, and keeping in beauty a long time. We have bloomed this species with seventy flowers on a plant at one time; sepals and petals greenish colour, based with chocolate; the lip a rich blue.

Z. rostratum.—A showy compact species from Demerara. A free-flowering plant; sepals and petals yellowish green; lip white striped with pink, and two inches across. This plant blooms three times a year, and lasts six weeks in perfection. It requires more heat and moisture than any of the other species.

INDEX.

	PAGE
INTRODUCTION	1
Directions for the Period of Growth	2
Water	3
On the cultivation of Tropical Orchids	4
Mode of Potting and the Materials to be used	5
Material for Terrestrial Orchids	7
Treatment of Fresh imported Plants	7
Advice to Collectors	8
Orchid Houses	11
Heating	13
Glazing	13
Ventilation	14
Shading	14
Cisterns	15
Treatment of Plants in flower and the best mode of protracting their Bloom	15
Treatment of Plants previously to being taken to a Flower Show	16
Remarks on preparing Orchids for travelling to a Flower Show	17
Treatment during the Time of Rest	20
Insects	22
Rot in Orchids	25
Spot in ditto	25
Propagation	27
Mode of producing Back Breaks	29
On the mode of making Baskets and the best wood for that purpose	29

	PAGE
Seedling Orchids	30
Orchids at present in cultivation	35
ACINETA	35
Barkerii	35
densa	36
Humboldtii	36
AERIDES	36
Affine	38
,, superbum	38
Crispum	38
,, Lindleyanum	39
,, pallidum	39
,, Warnerii	39
Cylindricum	39
Fieldingii	39
Larpentæ	39
Lobbii	39
maculosum	40
,, Schroderii	40
McMorlandii	40
nobile	40
odoratum	40
,, cornutum	41
,, majus	41
quinquevulnerum	41
,, album	41
roseum	41
,, superbum	42
suavissimum	42
Veitchii	42
virens	42
,, grandiflorum	42
,, superbum	42
Williamsii	43
AGANISIA	
pulchella	43

INDEX.

	PAGE
VARIEGATED ORCHIDS	43
ANŒCTOCHILUS	
argenteus	48
Bulleni	49
El-dorado	49
intermedius	49
javanicus	49
Lobbii	49
Lowii	49
,, virescens	50
Nevilleanus	50
"petola"	50
querceticolus	50
Roxburghii	50
Ruckerii	50
setaceous	51
,, cordatus	51
,, grandifolius	51
striatus	51
Veitchii	51
xanthophyllus	51
maculatus	52
ANGRÆCUM	
bilobum	52
caudatum	53
eburneum	53
,, superbum	53
,, virens	53
sesquipedale	53
ANGULOA	
Clowesii	54
,, macrantha	54
Ruckerii	54
uniflora	55
virginalis	55
ANSELLIA	
Africana	55
,, gigantea	55
ARPOPHYLLUM	
cardinale	56
giganteum	56
spicatum	56
BARKERIA	
elegans	57
melanocaulon	57
Lindleyanum	57
Skinneri	58
spectabilis	58
BLETIA	
campanulata	58
Shepherdii	59

	PAGE
BLETIA	
patula	59
BOLBOPHYLLUM	
barbigerum	59
Henshalli	59
saltatorium	59
BRASSAVOLA	
acaulis	60
Digbyana	60
glauca	60
venosa	60
BRASSIA	
Lanceana	61
Lawrenceana	61
maculata major	61
verrucosa	61
,, superba	61
Wrayii	61
BROUGHTONIA	
sanguinea	62
BURLINGTONIA	
amœna	62
candida	63
fragrans	63
Knowlesii	63
venusta	63
CALANTHE	
Dominii	64
furcata	64
masuca	64
,, grandiflora	65
Veitchii	65
veratrifolia	65
vestita rubra oculata	65
,, lutea	65
CAMAROTIS	
purpurea	66
CATTLEYA	
Aclandiæ	68
amabilis	68
amethystoglossa	68
bicolor	69
candida	69
citrina	69
crispa	69
,, superba	70
Edithiana	70
elegans	70
granulosa	70
guttata	70
,, Leopoldii	71

INDEX.

CATTLEYA
	PAGE
Harrisoniæ	71
,, violacea	71
hybrida	72
intermedia violacea	72
,, superba	72
labiata	72
,, atropurpurea	72
,, pallida	73
,, picta	73
Lemoniana	73
lobata	73
Loddigesii	73
marginata	74
maxima	74
McMorlandii	74
Mossiæ	74
,, aurantiaca	75
,, superba	75
pumila	75
quadricolor	75
Schilleriana	75
Skinneri	75
superba	76
violacea	76
Wagnerii	76
Walkeriana	76
Warnerii	77
Warcsewiczii	77
,, delicata	77

CHYSIS
aurea	78
bractescens	78
lævis	78
Limminghii	78

CŒLOGYNE
cristata	79
Cummingii	79
Gardneriana	79
Lowii	80
media	80
pandurata	80
plantaginia	80
speciosa	80

CORYANTHES
macrantha	81
maculata	81
speciosa	81

CYCNOCHES
barbatum	82
chlorochilum	82

CYCNOCHES
	PAGE
Loddigesii	82
pentadactylon	82
ventricosum	82

CYMBIDIUM
eburneum	83
giganteum	83
Mastersii	83
pendulum	83

CYPRIPEDIUM
barbatum	84
,, grandiflorum	84
,, superbum	84
biflorum	85
caudatum	85
,, roseum	85
Dayii	85
Farrieanum	85
hirsutissimum	85
insigne	85
,, Maulei	86
Lowii	86
purpuratum	86
Schlimii	86
Villosum	87

DENDROBIUM
aduncum	88
aggregatum majus	88
albo sanguineum	89
album	89
anosmum	89
Calceolaria	89
Cambridgeanum	90
chrysanthemum	90
chrysotoxum	90
clavatum	90
crepidatum	90
cretaceum	91
Dalhousianum	91
densiflorum	91
,, album	91
Devonianum	91
Falconerii	92
Farmerii	92
fimbriatum	92
,, oculatum	93
formosum	93
,, giganteum	93
Gibsonii	93
Heyneanum	93

INDEX.

	PAGE
DENDROBIUM	
Jenkinsii	94
lituiflorum	94
longicornu majus	94
Lowii	94
macrophyllum	95
,, giganteum	95
moniliforme	95
,, majus	95
moschatum	95
nobile	95
,, intermedium	96
,, pendulum	96
Paxtonii	96
Pierardii	96
,, latifolium	96
primulinum	96
pulchellum purpureum	97
sanguinolentum	97
taurinum	97
transparens	97
tortile	98
triadenium	98
Wallichianum	98
DENDROCHILUM	
filiforme	98
glumaceum	98
EPIDENDRUM	
alatum majus	99
atropurpureum, var. roseum	100
aurantiacum	100
aloifolium	100
bicornutum	100
cinnabarinum	101
crassifolium	101
Hanburyanum	101
macrochilum	101
,, roseum	101
maculatum grandiflorum	102
phœniceum	102
rhizophorum	102
Stamfordianum	102
verrucosum	102
vitellinum	102
,, majus	103
ERIOPSIS	
biloba	103
GALEANDRA	
Banerii	103

	PAGE
GALEANDRA	
cristata	104
Devoniana	104
GOODYERA	
discolor	105
Dominii	105
picta	106
pubescens	106
rubro-venia	106
GRAMMATOPHYLLUM	
Ellisii	107
speciosum	107
HOULLETIA	
Brocklehurstiana	108
odoratissima	108
HUNTLEYA	
marginata	108
meleagris	108
violacea	108
Wailesiæ	109
IONOPSIS	
paniculatus	109
LÆLIA	
acuminata	110
albida	110
anceps	110
autumnalis	110
Brysiana	110
Cinnabarina	111
elegans	111
,, Dayii	111
,, Warnerii	111
flava	111
furfuracea	112
grandis	112
Lindleyana	112
majalis	112
Maryanii	113
peduncularis	113
Perrinii	113
prestans	113
purpurata	113
,, Williamsii	114
Schilleriana	114
superbiens	114
xanthina	115
LÆLIOPSIS	
Domingensis	115
LEPTOTES	
bicolor	115
serrulata	115

INDEX.

LIMATODES
  rosea 116
LYCASTE
  cruenta 117
  Deppii 117
  Skinneri 117
    ,, alba . . 117
    ,, delicatissima . 117
    ,, rosea . . 118
    ,, superba . . 118
MILTONIA
  bicolor 118
  candida 118
  ,, grandiflora . . 119
  Clowesii major . . 119
  cuneata 119
  Karwinski . . . 119
  Morelii 119
  ,, atrorubens . . 119
  Regnelli 119
  spectabilis . . . 120
MORMODES
  crinitum 120
  luxatum 120
  pardinum 120
ODONTOGLOSSUM
  citrosmum . . . 121
  ,, roseum . . 121
  cordatum . . . 121
  coronarium . . . 122
  grande 122
  hastillabium . . . 122
  Insleayii . . . 122
  maculatum . . . 122
  membranaceum . . 122
  nævium 123
  ,, majus . . 123
  Pescatorei . . . 123
  Phalænopsis . . . 123
  pulchellum . . . 124
  Rossii 124
  triumphans . . . 124
  Uro-Skinneri . . . 124
  Warnerii 124
ONCIDIUM
  ampliatum majus . . 125
  Barkerii 125
  Batemanii . . . 125
  bifolium 125
  bicallosum . . . 126
  bicolor 126

ONCIDIUM
  Cavendishii . . . 126
  ciliatum 126
  crispum 126
  ,, grandiflorum . 127
  divaricatum . . . 127
  flexuosum 127
  Forbesii 127
  hæmatochilum . . 127
  incurvum . . . 127
  Lanceanum . . . 127
  leucochilum . . . 128
  longipes 128
  luridum guttatum . . 128
  oblongatum . . . 128
  ornithorynchum . . 128
  Papilio majus . . . 129
  phymatochilum . . 129
  pulchellum . . . 129
  pulvinatum . . . 129
  ,, majus . . 129
  roseum 130
  Sarcoides 130
  sessile 130
  sphacelatum majus . . 130
  unguiculatum . . . 130
  variegatum . . . 130
PAPHINIA
  cristata 131
  tigrina 131
PERISTERIA
  elata 132
  cerina 132
  guttata 132
PHAJUS
  albus 133
  grandifolius . . . 133
  Wallichii 133
PHALÆNOPSIS
  amabilis 136
  grandiflora . . . 136
  Lobbii 136
  rosea 137
  Schilleriana . . . 137
PLEOINE
  humilis 139
  lagenaria 139
  maculata 139
  Wallichiana . . . 139
PROMENÆA
  Rollissonii . . . 139

INDEX

	PAGE
PROMENÆA	
Stapelioides	139
RENANTHERA	
coccinea	140
SACCOLABIUM	
ampullaceum	141
Blumei	141
„ majus	141
curvifolium	141
furcatum	141
guttatum	142
„ giganteum	142
„ Holfordianum	142
miniatum	142
præmorsum	142
retusum	143
Wightianum	143
SCHOMBURGKIA	
crispa	143
Lyonsii	143
tibicina	144
SCUTICARIA	
Hadwenii	144
Steelii	144
SOBRALIA	
liliastra	145
macrantha	145
„ splendens	145
SOPHRONITIS	
cernua	146
grandiflora	146
violacea	146
STANHOPEA	
aurea	147
Devoniensis	147
grandiflora	147
insignis	147

	PAGE
STANHOPEA	
Martiana	147
oculata	147
tigrina	147
„ lutescens	147
TRICHOPILIA	
coccinea	148
crispa	148
picta	148
suavis	148
tortilis	149
UROPEDIUM	
Lindenii	149
VANDA	
Batemanii	150
cœrulea	150
cristata	150
gigantea	151
Lowii	151
insignis	151
Roxburghii	151
suavis	151
teres	152
tricolor	152
„ superba	152
violacea	152
WARRÆA	
cyanea	152
tricolor	153
ZYGOPETALUM	
brachypetalum	153
crinitum cœruleum	153
intermedium	153
Mackayii	153
maxillare	154
rostratum	154

THE END.

SELECT ORCHIDACEOUS PLANTS.

Part I.

By ROBERT WARNER.

Folio. Lovell Reeve.

The following is an extract from a notice of this Work in the *Gardeners' Chronicle*, July 5, 1862 :—

" This new contribution to our knowledge of Orchids will appear in ten quarterly parts, price 10s. 6d. each. The author states that 'having one of the largest amateur collections of Orchidaceous plants, and being in friendly correspondence with the principal growers throughout the country, he possesses ample opportunities for selecting the most interesting species and varieties for illustration. One object of the work will be to collect and disseminate information as to the best means of growing and flowering Orchidaceous plants. It is consequently intended to devote ample space to the details of cultivation; and the author will gladly avail himself of the assistance of those who are willing to communicate the results of their practice.' Mr. Williams, the author of the 'Orchid-Grower's Manual,' and one of the most skilful of cultivators, being associated in the undertaking, the public has the best possible guarantee that the foregoing promise will be kept. The species now figured are 1, Phalænopsis Schilleriana; 2, Cattleya amethystoglossa ; 3, Vandà insignis ; and 4, Cattleya Warscewiczii delicata. Of the three last noble plants the figures are worthy of Fitch's skilful pencil."

** The above Work can be sent by post from Paradise Nursery, Holloway, N.

SELECT ORCHIDACEOUS PLANTS.

By ROBERT WARNER, F.R.H.S.

The NOTES ON CULTURE by B. S. WILLIAMS,
Author of the "Orchid-Grower's Manual," and "Hints on the Cultivation of Ferns," assisted by some of the best Growers.

London : Lovell Reeve & Co.

Extract respecting this work from *Cottage Gardener*, July 22, 1862 :—

"This work promises to be excellent in every respect. The portraits of the Orchids in the first number, just published, are of life-size, and truthfully drawn and coloured. They are four in number—Phalænopsis Schilleriana, Cattleya amethystoglossa, Vanda insignis, and Cattleya Warscewiczii delicata. The work will embrace only those Orchids which are most beautiful and recently acquired, whether they be species or varieties, and thus will include 'real gems of the Orchid-house, which are rather ignored by those who look at the subject from a purely scientific point of view.' Yet there is enough of scientific description, given in language intelligible to every reader, sufficient to confirm him in his identifying the plant, which he may do, however, by the portraits, without any such aid. The directions for cultivating the Orchids portrayed are also full and good. In fact the work, judging from the specimen, will be such as might be expected from the authors, they being one of the largest amateur proprietors, and one of the best practical cultivators of Orchids in our times."

www.ingramcontent.com/pod-product-compliance
Lightning Source LLC
Chambersburg PA
CBHW022114160426
43197CB00009B/1020